GRAVITY, BLACK HOLES AND THE UNIVERSE

Iain Nicolson

A HALSTED PRESS BOOK

JOHN WILEY & SONS
New York

Also by Iain Nicolson

Astronomy: A Dictionary of Space and the Universe
The Road to the Stars
Black Holes in Space (with Patrick Moore)

Published in the U.S.A.
by Halsted Press, a Division of
John Wiley & Sons, Inc., New York.

ISBN 0-470-27111-6

Printed in Great Britain.

Contents

List of Illustrations

LINE ILLUSTRATIONS

ILLUSTRATIONS

PLATES

Introduction

What *is* gravity? This is a question which even Newton himself was careful not to attempt to answer.

We are all aware of the presence of gravity here on Earth; but, as we drag our weighty bodies out of bed in the morning, it is difficult for us to appreciate that gravitation is by far the weakest of the four "fundamental forces" known to physicists. Nevertheless, weak though it may be on the microscopic scale, gravity is the dominant force where large masses are concerned. It is gravity which holds us onto the surface of the earth, which chains the Moon in its orbit around the Earth, the Earth in its orbit around the Sun, and the Sun in its orbit around the centre of our star system. Gravity dominates the behaviour of the Universe as a whole, and will determine its future evolution.

The other three forces are the electromagnetic force which, like gravitation, operates across unlimited distances, and the strong and the weak nuclear interactions, which are effective only over tiny distances. Recent work has shown that the weak nuclear and the electromagnetic force appear to be no more than different manifestations of one underlying force, and there are hopeful indications that the strong nuclear force will soon be brought under the same umbrella. Gravitation, for the moment, remains out on a limb, the best available theory of gravitation—Einstein's General Theory of Relativity—treating it as an altogether different kind of influence from the other forces, despite Einstein's own efforts in later life. There are hopes that gravitation, too, will eventually be unified with the other forces but there are problems. Whereas the strengths of the others are such that their properties can be probed deeply by laboratory experiments, to distinguish between rival theories of gravity, one must examine large masses, intensely concentrated gravitational fields, and very large-scale phenomena. In effect, we must use an extremely large laboratory—the Universe itself.

In order to appreciate the rôle which gravitation has to play in the Universe as a whole, it is necessary to have an appreciation of the scale of the Universe and the nature of its principal constituents.

Everyday matter is made up of constituent atoms, each different chemical element having atoms with particular characteristics. In the simplest visualizations, atoms are considered to have a concentrated central nucleus made up of heavy particles known as protons, which have a positive electrical charge, and neutrons, which have no electrical charge; around the nucleus revolve a number of lighter, negatively charged, electrons. Nuclear physicists, in their ever-deeper investigations of subatomic and nuclear processes, have discovered that there exists a bewildering array of further "particles" of various kinds, many of them extremely short-lived; "fundamental" particles like protons and neutrons are themselves built up out of yet more basic particles, called quarks.

The heavy types of particle (including the proton and neutron) are known collectively as *baryons*, the "lightweight" particles (such as electrons) are known as *leptons*, and a group of particles of intermediate mass are termed *mesons*.

Atoms, and their constituent particles, are the basic components of the material Universe, their general behaviour and mutual interactions being governed by the forces of nature.

We live on the Earth, a planet travelling around the Sun on an elliptical orbit. Our nearest celestial neighbour is the Moon, the Earth's natural satellite, which travels around the Earth at a mean distance of some 384,000 kilometres (km), a distance roughly equal to ten circuits of the Earth's equator.

The Sun is a typical star. A self-luminous gaseous body, it has a radius about a hundred times that of the Earth and is about 330,000 times as massive as our planet. It produces energy by means of nuclear reactions going on in its central core in a process whereby matter is converted into energy in accordance with a principle which emerged from Einstein's Special Theory of Relativity (see Chapter 4). The mean distance between the Sun and the Earth is about 150 million km, and this distance defines a unit of measurement called the *astronomical unit* (AU).

It is often convenient, since light is the fastest-moving entity in the

Universe, to think of distances in terms of how long it would take for a ray of light to cross these distances. The velocity of light in empty space is about 300,000 kilometres per second (km/sec); at this speed a ray of light requires 1.3 seconds to reach us from the Moon, and 8.3 minutes to reach us from the Sun.

The Sun, together with nine known planets, various planetary satellites, other minor bodies and a certain amount of gas and dust—all of which travel around the Sun under its gravitational influence—comprise the Solar System. In order of distance from the Sun, the planets are Mercury, Venus, Earth, Mars, Jupiter, Saturn, Uranus, Neptune and Pluto. Their mean distances from the Sun range from 0.4AU for Mercury to just under 40AU for Pluto: a ray of light from the Sun requires about $5\frac{1}{2}$ hours to reach Pluto. Further out there is thought to lie a halo of comets, some of which make forays into the inner reaches of the Solar System.

The nearest star to us, apart from the Sun itself, is a faint red one known as Proxima Centauri. It is roughly 250,000 times further from us than is the Sun. A ray of light takes just over 4.2 years to reach us from that star; and this leads us to consider another unit of distance measurement, the *light-year.* A light-year is the distance travelled by a ray of light in one year, and it is equal to 9.46 million million km (written for convenience as 9.46×10^{12}km), or about 63,240AU.

The Sun is a member of a vast star system, of some 100 billion (10^{11}) stars, known as our Galaxy. The Galaxy is thought to be about 100,000 light-years in diameter (although some recent results suggest it may be significantly smaller) and consists of a central nucleus, where stars are relatively densely concentrated, surrounded by a flattened disc wherein stars are more sparsely distributed. The Sun is located in the disc, about 30,000 light-years from the galactic centre. The disc contains clouds of gas and dust out of which new stars continue to be formed, the stars and gas clouds being distributed in a roughly spiral pattern with respect to the nucleus. Distributed around the main body of the Galaxy is a halo of old stars, many of them concentrated in massive star clusters known as globular clusters.

Beyond the fringes of our own star system we can see billions of other galaxies, some spiral in shape, others elliptical, and still others quite irregular in shape. Our Galaxy is a member of a small cluster of galaxies known as the Local Group, the most conspicuous member of that group being the Andromeda Galaxy, which lies at a distance of some 2,200,000 light-years and is, incidentally, the most distant object which can be seen (under ideal conditions) with the naked eye. With telescopes we can see ordinary galaxies at distances in excess of 5 billion light-years, and there are other entities such as radio galaxies (which pour out prodigious quantities of radio waves) and quasars (see Chapter 10) which can be detected out to ranges possibly as great as 15 billion light-years. Such is the scale of the Universe which has so far been probed by Man.

As we shall see in Chapter 11, the evidence indicates that the Universe is expanding and that it probably originated in a great explosive event which took place, according to current estimates, some time between 10 and 20 billion years ago—the Big Bang. Whatever may have happened in those initial instants, gravity is the force which now controls the structure and destiny of the Universe. Whether or not the universe is finite in extent, whether it will continue to exist into an infinite future or come to a violent end, is governed by the gravitational interaction between all of its constituents.

If the broad canvas of the cosmos provides the greatest arena in which the effects of gravitation can be studied and tested, there are localized regions of space where gravity can overwhelm all else, namely black holes. Black holes are regions of intense gravitation formed, probably, by the collapse of matter; for example, the collapse of a massive star at the end of its life.

Although even light cannot escape from it, a black hole can still exert a gravitational influence on its surrounding, an influence which renders these bizarre objects detectable—at least, in principle. The mass of new astronomical information which has become available, particularly during the last fifteen years or so, largely as a result of our new ability to place instruments above the atmosphere, has led astronomers to believe that they may be detecting the effects of black holes in a wide variety of astrophysical contexts ranging from binary stars (pairs of stars in orbit around each other) to galaxies and quasars.

Theoretical developments cannot thrive when deprived of suitable experimental tests. After General Relativity, in the early part of the century, had successfully passed the few observational tests to which it could then be subjected, research interest in the subject lapsed—for all but a handful of esoteric phenomena, Newtonian ideas of gravitation were perfectly adequate for all gravitational calculations, including those involved in sending men to the Moon. But the recent upsurge of astronomical data, relating particularly to collapsed bodies like black holes and to the Universe as a whole, dramatically changed that situation. On the other hand, study of gravity has become an experimental science, with more sophisticated tests being devised and undertaken, and with the search for gravitational waves (see Chapter 12) now well under way. On the theoretical side, General Relativity itself is being explored more deeply, and extended, new theories are being examined, and the possibility of unifying gravity with the other forces is being investigated. Exciting new possibilities are being raised, epitomized by Professor Stephen Hawking's suggestion that black holes, hitherto regarded as immortal, may eventually explode! We seem to be poised on the brink of great new discoveries and theoretical advances; it is an exciting time.

In this book I have attempted first of all to show how our ideas on force, motion and gravity developed from earliest times to the triumphant syntheses achieved by Newton and by Einstein. I have explored in more detail the theoretical and observational aspects of those two areas of greatest contemporary interest in which gravity plays a significant rôle—black holes, with their many bizarre properties, and cosmology, the study of the structure and evolution of the Universe. Finally, I have tried to look at the present and future of General Relativity and alternative theories of gravitation, and have examined such questions as whether or not the strength of gravity varies with time and the extent to which new experimental tests may be expected to improve and alter our understanding of this most elusive, yet all-pervading force.

PART ONE
CHANGING VIEWS

I
Force and Motion in the Ancient World

To early Man the world must have been a bewildering place, full of phenomena beyond his control or comprehension. The great natural forces of wind and water—storms, thunder, lightning, earthquakes and floods—seemed to be subject to the whims of supernatural beings. Yet there were also regular patterns of events which, as human societies became more sophisticated, gradually became apparent. The most obvious cycle was provided by the daily rising and setting of the Sun, and the Sun itself, the provider of light and heat, was deified in most early cultures. The Moon provided light by night, and its changing cycle of phases formed the basis of another useful unit of time measurement, the month. It, too, was usually deified, but in contrast to the Sun (which was generally recognized as having male characteristics), the Moon was usually ascribed female qualities.

The longer-term cycle of the seasons was of vital importance, particularly with the development of agriculture. The stars were constant, unchanging in their patterns, or constellations, but the Sun and Moon moved relative to the stars, as did a number of "wandering stars", the five planets whose existence had been known since prehistoric times. The changing position of the Moon was readily visible from night to night, and the fact that the Sun moved through the stars could be established by observing that different stars were visible close to the horizon just before sunrise or just after sunset at different times of the year. The motions of the planets were less obvious, but nevertheless readily apparent to the early sky-watchers.

There was a regular and ceaseless pattern to events in the heavens, the Sun, Moon and planets moving according to repetitive cycles, while the stars remained immutably in their appointed positions. Here on Earth, things were different. Changes occurred, often rapidly and unpredictably. Heavy objects fell down, flames rose upwards, liquids spread out and flowed, while the air moved around us, invisibly, but tangibly. Who or what controlled these motions? As generations of enquiring minds pondered such questions, mankind embarked upon a quest for knowledge and understanding of the Universe, for natural explanations of phenomena—a search for order in the world, for patterns and structures, and for the fundamental forces which govern the behaviour of the Universe and all that it contains.

All early cosmologies placed the Earth at the centre of the Universe. The Earth was usually envisaged as being flat or lens-shaped, and the sky was regarded as a dome set above it. Deities were to be found in profusion. For example, in the ancient Egyptian view, the sky was the body of the goddess Nut, supporting herself above the Earth-god Qeb, while the Sun was born each morning and sailed across the heavens on a barge. However, more sophisticated ideas were to follow. The Ionian civilization, which flourished around 700 BC on the shores of the Aegean Sea, developed the idea that the Earth lay at the centre of a spherical Universe, and later—perhaps because of a developing sense of symmetry—there came the notion that the Earth itself might be a sphere. The spherical Earth lay at the centre of a spherical cosmos.

The Greeks were great geometers, and had a well developed sense of symmetry and geometric perfection. Just as the circle was the most perfect curve, for it had no beginning and no end, and every point on it was equidistant from the centre, so the sphere was the most perfect solid. The idea evolved that in the perfect and unchanging heavens only perfect shapes and motions were possible.

How did the Sun, Moon, stars and planets move? The Pythagoreans tried to treat their motions as being circular, but in a different approach Eudoxos (*c.*408–355 BC) devised a system of Earth-centred concentric spheres to account for the behaviour of celestial bodies. The stars were fixed to the outermost sphere which rotated around the Earth once per day, so accounting for their rising, east-to-west motion, and setting. Three spheres each were required to explain the movements of the Sun and the Moon, while each of the planets needed four spheres in relative motion to describe their behaviour. Consider Jupiter, for example (fig. 1). It shared in the daily motion of the heavens, and the outermost of its set of spheres took care of that aspect. Carried inside that sphere was another which was inclined to the axis of the first by an angle of about 23 degrees and which revolved in just under twelve years, so accounting for Jupiter's slow change of position relative to the background stars. The two other spheres were necessary to cater for a curious aspect of planetary behaviour whereby the planets from time to time stop and then reverse their directions of motion for a while before resuming their normal west-to-east motion relative to the background stars. As a result of this behaviour the planets Mars, Jupiter and Saturn were seen to trace out "loops" in the sky at regular intervals, while Mercury and Venus also moved in rather disconcerting fashion. In all, Eudoxos required a total of 27 moving spheres.

These ideas were taken up by Aristotle (384–322 BC), the most influential of the great Greek philosophers. He elaborated the theory

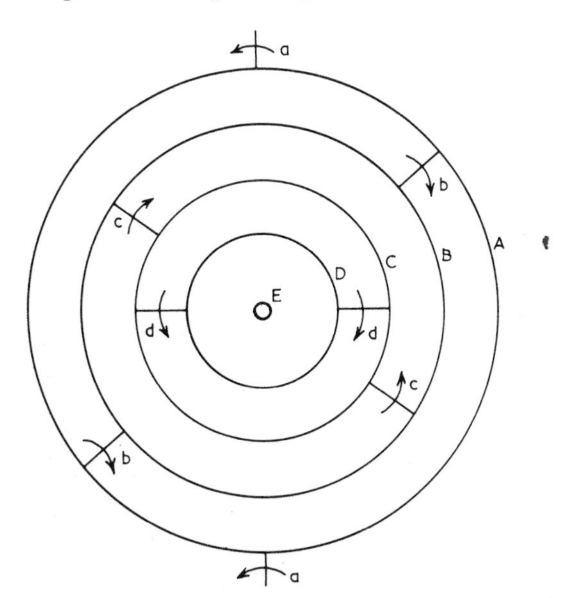

Fig. 1 *The planetary spheres of Eudoxos.* To account for the motion of a planet such as Jupiter, Eudoxos required four rotating spheres. Sphere A rotated in 24 hours to account for the daily motion of the heavens; while sphere B rotated, about an axis inclined to the first by some 23.5°, in an anticlockwise direction in a period of about 12 years, so accounting for Jupiter's motion relative to the stars. Spheres C and D revolved in opposite directions over a period of 13 months to account for the retrograde loops traced out by the planet at 13-month intervals.

to such an extent that 55 moving spheres were required to account for all the celestial motions. In his system the Earth was unmoving, spherical, and firmly located at the centre of the Universe.

Aristotle was a notable observer of nature. He proved that the Earth was spherical by noting the curved shape of the Earth's shadow on the Moon during a lunar eclipse and by reference to the fact that the stars which could be seen altered as one moved north or south. The vanishing of ships over the horizon was explained in similar fashion. He made notable contributions to the development of thought in many varied directions—political, economic, ethical and metaphysical; in addition, he may justly be claimed to be the founder of biology.

Aristotle's elements and mechanics

Aristotelian mechanics was firmly rooted in the concept of a fixed Earth located at the centre of the Universe, and his views became an entrenched foundation of philosophical and religious dogma which was not overthrown until the seventeenth century. His mechanics embraced the ideas of *natural motion* and *natural place.* Following the views expressed in the previous century by Empedocles (*c.*484–424 BC), Aristotle developed the idea that bodies were composed of four basic elements—air, earth, fire, and water. Earth was absolutely heavy and fire was absolutely light, while air and water were intermediate in this respect. The natural place for earth was at the centre of the Earth, which was the geometric centre of the Universe; consequently, the natural motion for earthy substances is to fall directly towards the centre of the Earth, as heavy objects are seen to do. The natural motion for fire was to rise straight up to its natural place in the firmament. The natural place for water was above the Earth, and the natural place for air was above water but below fire. There was no concept of a force of gravity in these ideas, but the natural place for heavy bodies was the centre of the Universe which was also the centre of the Earth.

The motion of an arrow, or of a stone thrown in the air, was not natural; rather it was forced, or "violent" motion. This kind of motion could take place only when a *force* was acting to start the body on its way, and to keep it going. When the motive force was exhausted, the body would seek its natural place. Thus a stone hurled upwards would continue to rise for as long as the force kept it going; when the force ran out, it would fall to Earth along a straight-line path, seeking its natural place. According to Aristotle, all motion in the vicinity of the Earth took place along straight lines. The absurd conclusions to which this doctrine led are well illustrated by the case of the flight of an arrow. If fired into the air at an angle to the horizontal, it would continue to move in that direction (in a straight line) for as long as the force from the bow continued to push it along. When the motive force ceased, the arrow would plummet vertically down to Earth. We can easily convince ourselves by the simple expedient of throwing a pebble that the flight of a projectile follows a curved path (a parabola) rather than two successive straight lines. One would have thought that simple observation (which he encouraged) would have shown the inadequacy of Aristotle's theory of motion, but such was the power of his complete system of natural philosophy that no acceptable alternative view emerged for well over a thousand years. (See inset to fig. 9, page 40.)

The heavens were separate and different from the Earth. On the Earth, the four elements were corruptible: things came into being, flourished, and decayed. But in the heavens nothing changed; therefore celestial bodies had to be made of a fifth element, pure and incorruptible—the *aether*. The only celestial body which showed a changing appearance was the Moon, with its blotchy appearance and changing phases. Aristotelian doctrine suggested that the sphere of the Moon (i.e., the sphere on which the Moon was located as it turned around the Earth) marked the boundary between the incorruptible heavens and the base and changeable world of the Earth. Beyond the sphere of the Moon all was perfection.

The circle represented the natural motion of the perfect celestial bodies, while both natural and violent motion in the vicinity of the Earth took place along straight lines. Clearly the Earth itself could not move, either on its axis or around the Sun, for that would not be the natural motion of earthy matter. He refuted suggestions that the Earth might spin on its axis by pointing out that when a heavy object is thrown straight up it returns to its starting point; likewise a weight dropped from a tower lands at the foot of the tower. If the Earth were rotating, the stone would land well away from the foot of the tower (using present-day values for the rate of fall and the size of the Earth, if Aristotle's argument were correct a stone falling from an 80-metre tower would land about two kilometres west of the tower!). By the same token, there would be a terrible wind blowing across the surface of the Earth if it were to spin in this patently unnatural fashion.

Aristotle also argued against motion around the Sun on the grounds that, if the Earth did move in this way, it would be nearer to different parts of the sphere of stars at different times (fig. 2), and the apparent positions of stars would be seen to change; for example, two stars should be observed to move further apart as the Earth approached them, and to draw closer together as the Earth receded. As this parallax effect was not observed, the Earth could not move. In fact this kind of positional change does occur but can be detected only by careful and precise telescopic observations, simply because the stars are so much further away than could have been conceived in Aristotle's time.

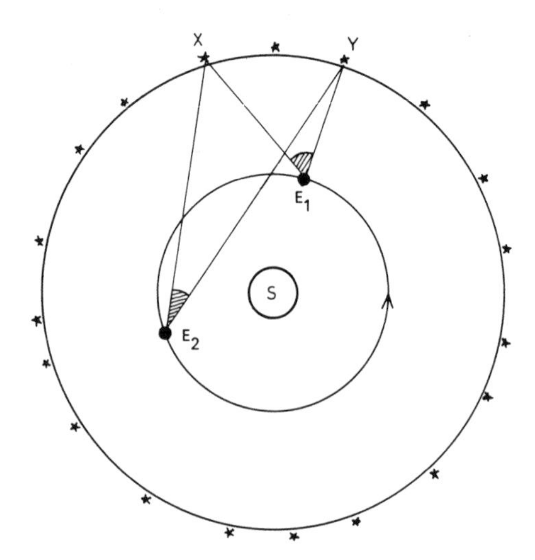

Fig. 2 *Refuting the motion of the Earth.* One argument used by the Greeks against the possibility that the Earth might move was that, if it did, then as it moved, say, from E_1 to E_2 the angle between the stars X and Y would change appreciably. Since this was not observed, the Earth could not move.

All motion required an active force. Natural motion near the Earth was due to the inherent qualities of heaviness ("gravity") or lightness ("levity"), which caused bodies to move down or up. The natural circular motion of the celestial spheres was considered to be due to a divine prime mover.

While it was easy to see the connection between force and motion where lifting, pulling or pushing were concerned, the forced motion of bodies through the air posed problems. If a horse were pulling a cart, motion would take place only while the horse continued to pull; if the horse stopped the cart too would stop. With the flight of an arrow from a bow, how was the force communicated to the arrow? Aristotle concluded that it must be transmitted *via* successive layers of air from the bow to the arrow. An earlier suggestion, due to Plato, was that the air in front parted, moved around behind the arrow, and gave it a push.

The medium through which the arrow was travelling transmitted the force, but also resisted the motion. For motion to take place, the force had to exceed the resistance; uniform motion required an appropriate balance between the two opposing factors. The speed of a moving body was directly proportional to the force (the greater the force, the faster the motion) and inversely proportional to the resistance (the greater the resistance, the slower the motion). The concept of resistance was an important one in Aristotelian mechanics.

Following on this line of argument, Aristotle claimed to show the impossibility of a vacuum. If a body achieves a given speed with a given force in a dense medium, it will achieve a higher speed in a less dense one. In a vacuum there could be no resistance; since speed is inversely proportional to resistance the speed of bodies in empty space would become infinite. Since such a conclusion was clearly absurd, truly empty space could not exist; instead, all space must be filled with some kind of medium. The concept of such an all-pervading aether was to haunt scientific thought until the twentieth century.

In Aristotle's view, bodies of the same size, shape and weight should fall at the same rate since the force ("gravity") and the resistance on each would be the same. However, if two bodies of the same size and shape, but of different weights, were released from a height, the heavier should move faster, and reach the ground sooner, because, although the resistance acting on both weights would be the same, the force (weight) would be greater in the case of the heavier object. The acceleration of falling bodies was accounted for by the heaviness of a body being enhanced as it approached closer to its natural place.

It was, then, a clear conclusion of Aristotelian mechanics

that—provided similar-sized bodies were being discussed—*the heavier the body, the faster it should fall.* This was a view which persisted until the scientific revolution nearly two millennia later, and even today the belief is quite common that heavier bodies fall faster than light ones—it seems so reasonable.

The development of planetary theory

There were those who opposed the idea of a geocentric Universe. Philolaos, a Pythagorean who lived in the latter part of the fifth century BC, had suggested that the Earth travelled daily round a central fire (not the Sun), for reasons connected with the Pythagorean obsession with numbers. Heracleides (*c*.390–310 BC) proposed a scheme whereby Mercury and Venus travelled around the Sun, while the Sun and the other planets moved around the Earth. Aristarchus (*c*.310–230 BC) went so far as to suggest that all the planets, and the Earth itself, travelled around the Sun; however,

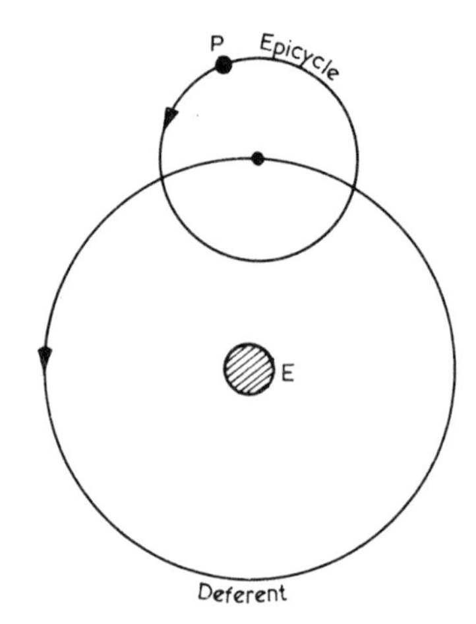

Fig. 3 *Epicycle and deferent.* This is one of a number of devices used in the Ptolemaic system. The planet P was supposed to move on a small circle, the epicycle, while the centre of that circle moved around the Earth on a larger circle, the deferent.

this idea was well before its time, and was rejected for a variety of reasons including its incompatibility with Aristotelian mechanics (the natural place for heavy bodies being the centre of the Earth) and the absence of any parallax effect.

Ptolemy (*c*. AD 100–165) argued that, since the centre of the

Universe is the natural home of earthy material, the Earth must be there. If it were not then, being heavier than any of the objects on its surface, it would be falling towards the centre of the cosmos much faster than anything else, leaving objects on it, such as trees, animals and people, far behind, floating in the air!

In attempts to make better calculations of planetary motions, the concept of concentric spheres was abandoned in favour of uniform circular motion. In order to fit the observed movements of the planets into such a scheme it was necessary to invent various devices. The best known is the *epicycle* (fig. 3). The planet was assumed to move on this small circle at a uniform rate, while the centre of the epicycle moved at a uniform rate on another circle, the *deferent*, which was centred on the Earth. A prime reason for the introduction of these devices was to account for the retrograde (backward) motion of the planets when, from time to time, they traced out loops in the sky. Today we know that this effect is simply an apparent motion experienced when the Earth is "overtaking" one of the outer planets (fig. 4). To use an everyday analogy, we know that as we overtake a vehicle in the slow lane it appears to move backwards relative to the distant background.

Other devices, too, were necessary to account for apparently nonuniform motions in the sky. The *eccentric*, a device invented by Apollonius (*c.*262–190 BC), was a circle with its centre displaced from the Earth, so that uniform motion on this circle would appear nonuniform when seen from the Earth. Ptolemy, who summarized and synthesized the Greek view of the Universe in his book the *Almagest*, added another device of his own—the *equant*, by means of which a planet was allowed to perform nonuniform motion on its circle provided that there was one point (not the Earth) within the circle from which a hypothetical observer would see motion which would *appear* to be uniform! Such a concept surely was straining the ideal concept, of uniform motion on circular paths, beyond all reasonable bounds.

By a judicious selection and combination of these devices, each planet having its own separate scheme, Ptolemy was able to devise a system of planetary motion which enabled the future positions of planets to be calculated with surprising accuracy. But it was a complicated and cumbersome system which tarnished the concept of uniform circular motion and was far removed from the simple

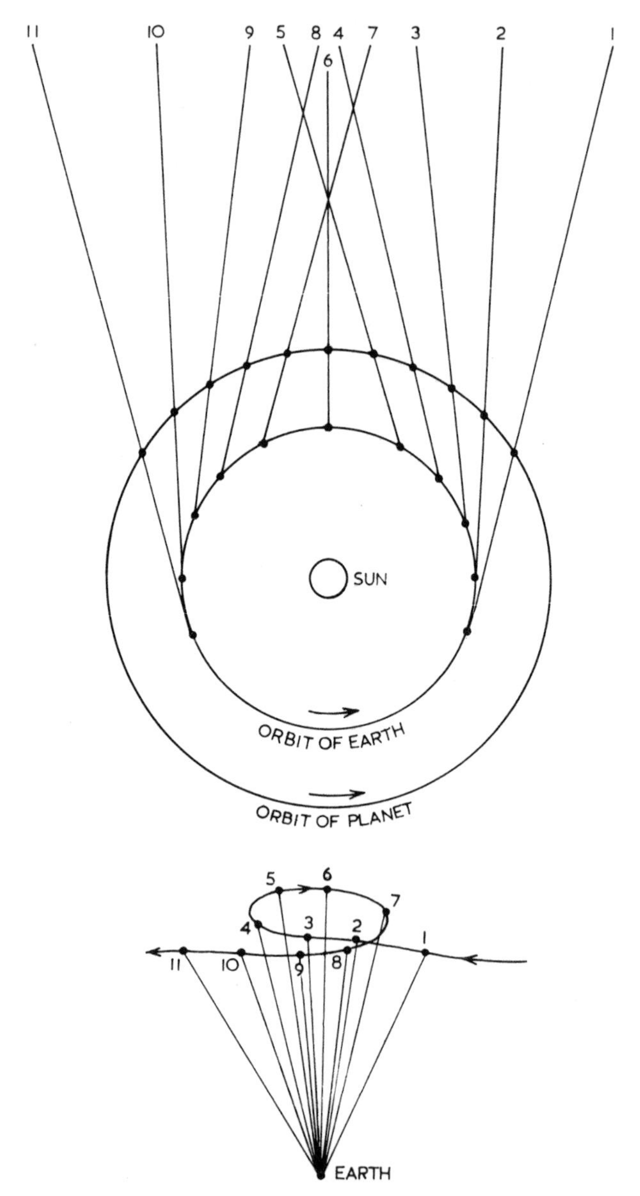

Fig. 4 *Planetary loops.* The upper part of the diagram shows how, as the Earth overtakes a slower-moving outer planet, the line of sight from the Earth to that planet changes direction. The lower part shows how the changing line of sight results in the planet halting in its west-to-east motion, running backwards (retrograde) for a time, before resuming its forward (direct) motion, thus tracing out a loop on the sky.

geometric perfection which earlier generations of philosophers had envisaged.

The Ptolemaic system of the Universe and the Aristotelian view of force and motion, wherein force was necessary to maintain motion and material objects were composed of elements which sought their natural places, were transmitted throughout the Middle East and Europe. Although the view of the world which they gave was a beautifully self-consistent one, it was glaringly at variance with experience and observation in a number of respects. Nevertheless, it remained largely unchallenged for well over a thousand years: that is the measure of the achievement of the Greek philosophers.

2
Changing Views of the Universe

Together with the principles of Aristotelian physics, the Ptolemaic system of a geocentric Universe was handed down with little significant change *via* the Middle East and thence, through Spain, to mediaeval Europe. However, it would be wrong to suggest that Ptolemy's system was accepted uncritically. For example, Alfonso X, king of Léon and Castile in Spain from 1252 until 1284 and instigator of the Alfonsine Tables of eclipses and planetary positions, when first introduced to the complexities of that system is alleged to have remarked, "If the Lord Almighty had consulted me before embarking on the Creation, I should have recommended something simpler." Undoubtedly, one of the main objections to the Ptolemaic system was its elaborate and ponderous nature.

Reasoned objections began to be voiced. The great French philosopher Nicolas of Oresme (1320–1382) pointed out that it was simpler to imagine the Earth rotating than the great sphere of stars spinning round, and he was by no means the first to do so. He did not go so far as to accept the physical *reality* of the Earth's rotation, but he did suggest that the outermost sphere might be surrounded by an infinite empty space. These ideas were taken further by Nicholas of Cusa (1401–1464), a German cardinal and philosopher, and an advocate of the experimental approach. In his opinion the Earth really did rotate on its axis, and the Universe might be infinite in extent with *no* fixed and immovable centre.

The work of Nicolaus Copernicus (1473–1543) initiated the revolution which was to overthrow the Ptolemaic system and, with it, the whole framework of Aristotelian mechanics. Born in Torun, in Poland, Copernicus became acquainted while a student with the idea that the Earth might move. He became convinced that the observed motions of the celestial bodies would be explained in a more satisfactory fashion by assuming both that the Earth rotated on its axis and that the Earth and planets moved around the Sun, which was located at the centre of the Universe. In his system the order of distance of the planets from the Sun was as follows: Mercury, Venus, Earth (and Moon), Mars, Jupiter, Saturn. Beyond Saturn lay the sphere of the fixed stars (fig. 5).

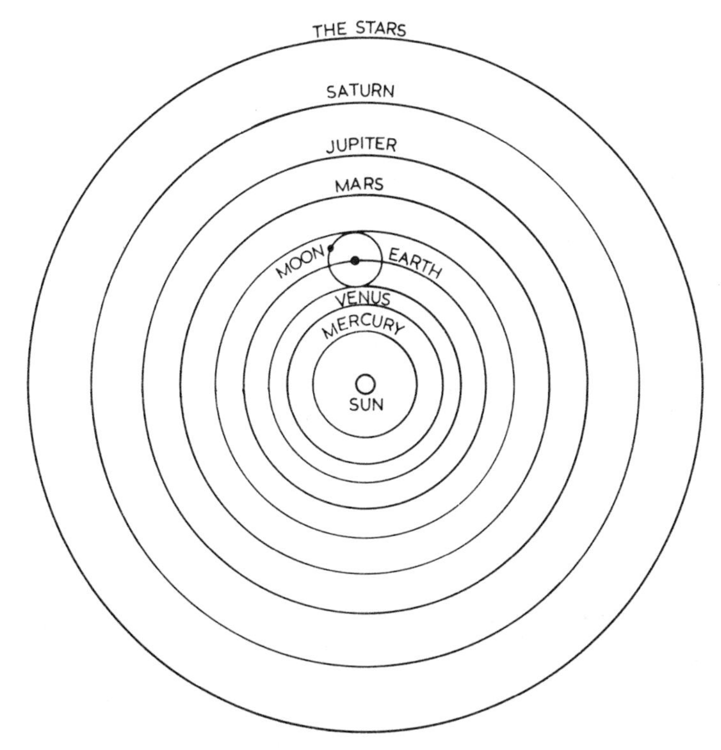

Fig. 5 *The Copernican system.* In the heliocentric system of Copernicus, the planets pursued circular orbits around the Sun in the order shown here. The Moon moved around the Earth; the sphere of fixed stars lay far beyond the orbit of Saturn.

The Copernican heliocentric system was in essence simpler than the Ptolemaic one. The rotation of the Earth on its axis removed the need for the daily motion around the Earth of the sphere of stars, and of every celestial body, while the motion of the Earth around the Sun accounted for both the apparent annual motion of the Sun around the celestial sphere (fig. 6) and the retrograde loops of the planets (see fig. 4, p. 24). However, Copernicus adhered rigidly to the idea of perfect uniform circular motion in the heavens and, because of this, still found it necessary to use epicycles and eccentrics to obtain better than a rough fit to the observed motions of the planets: in the end, according to Copernicus, his system required 34 circular motions, considerably fewer than any version of the Ptolemaic system; nevertheless, it has been argued that the Copernican system *actually* used 48 motions, while one version of the Ptolemaic system made use of only 40. The doctrine that uniform circular motion was the only type of motion appropriate for heavenly bodies still lay heavily on planetary theory—indeed, Copernicus felt that his system adhered more truly to this doctrine than did the Ptolemaic model because Ptolemy's system incorporated the device of the equant, which involved nonuniform motion.

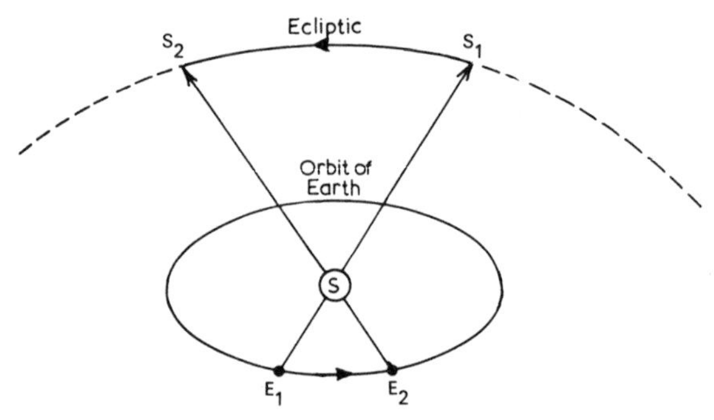

Fig. 6 *Apparent motion of the Sun.* As the Earth moves around the Sun from E_1 to E_2, so the Sun changes its position against the background stars from S_1 to S_2. In the course of a year the Earth completes one circuit of the Sun, while the Sun appears to trace out a circle, the *ecliptic*, around the celestial sphere.

Many of the objections raised against the heliocentric view were the same as those which had been used in Greek times. The concept was wholly at variance with the Aristotelian doctrine of natural place and natural motion. Another objection concerned the absence of parallax effects. In answer Copernicus postulated, correctly, that the sphere of stars was so vast in comparison with the orbit of the Earth that no parallax could be observed. In his own words, "the distance of the Earth to the Sun is as nothing to the height of the firmament".

Although Copernicus had circulated a manuscript containing an outline of his ideas, it was not until 1540 that he was persuaded by Tiedemann Giese and Georg Joachim Rheticus to publish his work in full. Rheticus entrusted the publication of the Copernican theory to Andreas Osiander, a Protestant German cleric who could not accept the idea that the Earth *actually* moved, as it was apparently contrary to scripture. Consequently, when *De revolutionibus orbium coelestium* was published in 1543, the year of Copernicus' death, Osiander wrote an anonymous preface stating that the idea of a moving earth was a mathematical device to simplify and improve the calculation of planetary positions, rather than a physical fact. It is clear from the text of the book, however, that Copernicus himself truly believed that the Earth did move.

De revolutionibus was dedicated to Pope Paul III. It caused considerable interest and debate but, at that time, no official opposition from the Catholic Church. The Protestants were more strongly incensed, Martin Luther declaring: "The fool will turn the whole of the science of astronomy upside down. But, as Holy Writ

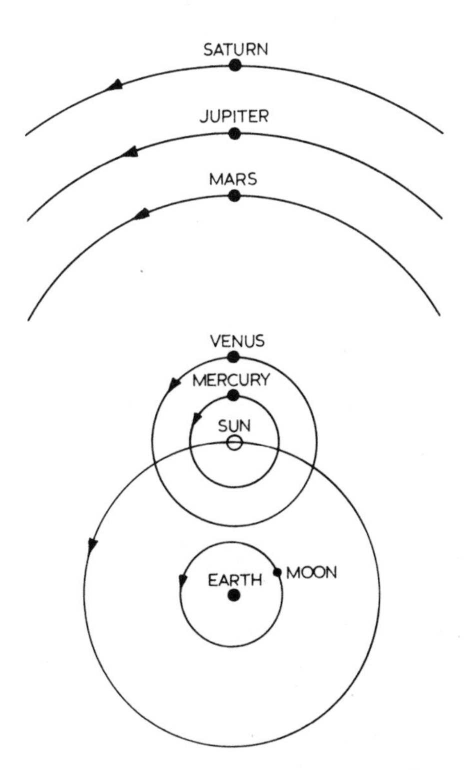

Fig. 7 *The Tychonic system.* According to Tycho Brahe, the Sun and Moon travelled around the Earth on circular paths, while the planets travelled, also on circular paths, around the sun.

declares, it was the Sun and not the Earth which Joshua commanded to stand still." At that time there was no hint of the reaction which was later to hamper the work of avowed Copernicans like Galileo.

The idea of a heliocentric Universe and a moving Earth began to find favour among scholars, and in England the theory was given firm support, notably by Thomas Digges (*c.*1545–1595). In his book, *A perfit description of the coelestiall orbes*, published in 1576, Digges gave an English translation of much of *De revolutionibus* and linked the heliocentric theory with his own ideas of an infinite Universe filled with stars. In 1583 the Dominican Friar Giordano Bruno visited England and learned of Copernican theory. His espousal of Copernicanism and of the concept of an infinite stellar Universe became linked in the minds of Catholic churchmen with Bruno's generally heretical attitude towards the Church and its government. Bruno was burned at the stake for his heresies in 1600, and it was largely as a result of his activities that the Catholic Church was led to condemn the Copernican theory. Over seventy years after its publication, *De revolutionibus* was placed on the Index of Prohibited Books in 1616.

By no means did every astronomer of the period find the Copernican viewpoint attractive. Tycho Brahe (1546–1601), a Dane, who was without doubt the finest observational astronomer of his time, could not accept the idea of a moving Earth and instead advanced his own hypothesis (fig. 7). He retained the Earth firmly at the centre of the Universe and adhered to the Ptolemaic concept of the stars being fixed to a sphere which rotated around the Earth once per day. However, he considered that the planets moved around the Sun in circular paths, but that the Sun itself moved round the Earth, a viewpoint remarkably similar to the one which had been advanced by Heracleides nearly two thousand years earlier. Despite its undoubted attractions, Tycho's theory did not receive much support.

His precise observations of the motions of the planets were to provide the data necessary for the next step in planetary theory, as we shall see later. Among his other noteworthy observations was his study of the supernova of 1572, a star which suddenly appeared in the constellation of Cassiopeia, flaring up to become as bright as the planet Venus, then fading from view after about sixteen months. His careful observations revealed no parallax, which meant that this object must lie well beyond the sphere of the Moon. Since it did not share in the planetary motion either, Tycho concluded that it must lie in the sphere of stars, so contradicting the long-held Aristotelian view that the sphere of fixed stars was unchanging. Five years later he examined a comet and showed that it moved in an orbit around the Sun and was further away than Venus, thus confounding the Aristotelian doctrine—previously held by Tycho himself—that comets were atmospheric phenomena. Yet, although these observations contradicted the traditional view, he could not abandon the idea that it was impossible for the ponderous Earth to move.

The most famous champion of the Copernican system was the Italian professor Galileo Galilei (1564–1642), who is perhaps best known for his use of the telescope, although his contributions to mechanics were of vital importance in the march forward towards a consistent theory of mechanics and a theory of gravity. He became acquainted with Copernican ideas, but was at first unconvinced. By 1604, however, he had devised mathematical arguments which persuaded him that the Earth did move, and in that year, too, a bright nova appeared which Galileo publicly declared to lie beyond the sphere of the Moon, a view which opposed the Aristotelian doctrine of the immutability of the heavens.

In 1609 he heard of the invention of the telescope, traditionally ascribed to the Dutch optician Hans Lippershey, and proceeded to make his own instruments, one of which magnified thirty times. In the winter of that year Galileo turned his telescopes to the skies, and what he saw was astounding and revolutionary. Within a few months of observation he made discoveries which altered the whole of Man's conception of the cosmos. He found much which was contrary to Aristotelian doctrine, including evidence to favour the Copernican system. He saw that the Moon had mountains, valleys, and other features which showed that it was like the Earth, a world, and not a smooth, perfect and changeless orb of celestial material. He found that there were innumerable stars too faint to be seen without optical aid, and that the Milky Way was made up of a mass of faint stars, rather than being an atmospheric phenomenon as Aristotle had thought. He noted also that, while the planets were magnified into visible discs, stars still looked like points of light when seen through the telescope: this suggested that the stars must lie at very great distances.

The Sun was shown (although not for the first time) to have spots on its surface, and from these spots Galileo deduced the rotation of the Sun. The Sun was not the perfect and unblemished aethereal body it had been supposed to be; moreover, if it spun on its axis, surely the Earth could do likewise. Venus was shown to pass through a cycle of phases, and this was not possible on the Ptolemaic model. Perhaps the most important observations published in his *Sidereus Nuncius* ("The Starry Messenger") in 1610 concerned the four satellites which he had seen revolving around the planet Jupiter. This showed that the Earth was not the only centre of motion in the Universe and rendered more plausible the Copernican hypothesis that the Earth itself moved around the Sun.

Of themselves, Galileo's observations did not constitute proof of the heliocentric theory—indeed, all his results could have been accommodated within a Tychonic system—but they were consistent with the Copernican point of view, and did much to demolish the old distinction between the heavens and the Earth.

The other key figure in the establishment of the heliocentric system was Johannes Kepler (1571–1630). Kepler was born at Weil, in the German province of Württemberg, to less than happy circumstances, and his life was filled with personal trials. Nevertheless, although in many respects his ideas were more in tune with the ancient mystical world, it was he who dealt the conclusive death blow to the geocentric theory and the dogma of uniform circular motion in the heavens.

In his early work he developed a heliocentric system of the planets whereby the planetary distances were determined by reference to the shapes of fundamental geometric solids; although this aspect of his work is considered today to be largely worthless (and contemporaries such as Galileo were scathing in their criticism of it), its publication in 1596 in his book, the 24-word title of which is usually abbreviated to *Mysterium Cosmographicum*, attracted the interest of Tycho, who invited Kepler to join him at Prague to help analyse Tycho's observations of the motions of the planets.

In 1600 Kepler set to work on the motion of the planet Mars. He tried all kinds of combinations of epicycles, deferents, eccentrics and equants in an attempt to fit the observed motion of the planet, and was able by this means to get agreement to an accuracy of 8 minutes of angular measurement (a little over one eighth of a degree, or a quarter of the apparent diameter of the Moon in the sky). This was

Aristotle's theory of projectile motion. Gunner firing a cannon whose elevation has been calculated using a chequered rule fitted with plumb bobs. According to Aristotelian physics no body could undertake more than one motion at a time, and the path of the projectile therefore had to consist of two separate straight lines. From Daniele Santbech, *Problematum Astronomicorum*, Basle, 1561. (Ann Ronan Picture Library.)

Nicolaus Copernicus (1473–1543), the Polish astronomer whose heliocentric system of the Universe, published in 1543, initiated a revolution in our conception of the Universe. (BBC Hulton Picture Library.)

Johannes Kepler (1571–1630), the German astronomer who, from his analysis of Tycho Brahe's precise measurement of the motions of the planets, was able to show that the Earth and the planets travel round the Sun in elliptical orbits. His laws of planetary motion were a vital link in the line of reasoning which led Newton to his law of universal gravitation. (BBC Hulton Picture Library.)

Galileo Galilei (1564–1642), the Italian physicist and astronomer who was an outspoken proponent of the Copernican system. His telescopic observations dramatically altered our view of the celestial bodies, and his experiments contributed greatly to the overthrowing of Aristotelian physics, while providing the basis for the subsequent development of Newton's laws of motion. (BBC Hulton Picture Library.)

not good enough for Kepler: in his own words, "these 8′ alone lead the way toward the complete reformation of astronomy". Finally, after some seventy attempts, he stumbled upon the answer. The orbit of Mars had to be an *ellipse.* This was a conclusion which he accepted with considerable reluctance for, like his contemporaries, he was thoroughly steeped in the tradition of perfect circular motion.

His discoveries were published in 1609 in *Astronomia Nova,* and in later years he extended his work to include even the satellites of Jupiter, publishing his results between 1619 and 1621. Amid the great mass of useful information and confused speculation lay three crucial laws (the first two were published in *Astronomia Nova,* the third in *Harmoniae Mundi* (1619)), known today as Kepler's laws of planetary motion. In present-day terminology, they are as follows:

First law: The path of each planet around the Sun is an ellipse, the Sun being located at one focus of the ellipse (fig. 8);

Second law: The radius vector (i.e., the line joining the planet to the Sun) sweeps out equal areas of space in equal times, so that in their elliptical orbits planets move faster when closer to the Sun than they do when further away;

Third law: The square of a planet's periodic time (the time required for it to complete one orbit of the Sun) is directly proportional to

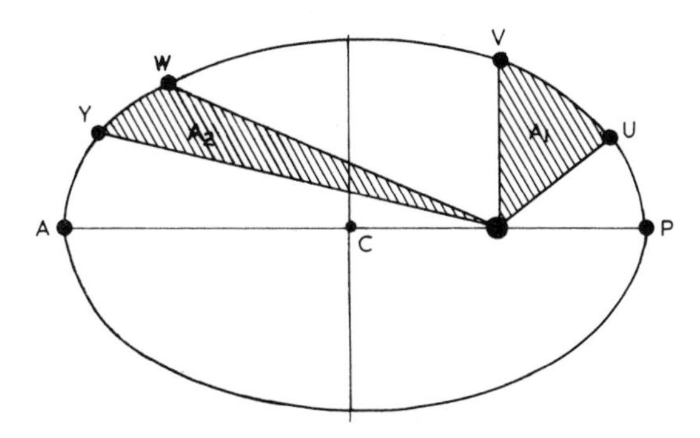

Fig. 8 *Kepler's laws of planetary motion.* The orbit of a planet is an ellipse with the Sun located at one focus of that ellipse. AP is the major axis; AC is the semimajor axis, or "mean distance from the Sun" in Kepler's third law. The point of closest approach to the Sun (P) is termed perihelion while the point of greatest distance (A) is termed aphelion. According to Kepler's second law, if the planet moves from U to V in a given period of time, it will move from W to Y in the same period of time, so that area A_1 = area A_2.

the cube of its mean distance from the Sun (by "mean distance" we refer to the semimajor axis of the ellipse).

The third law implies that the mean distance of a body from the Sun can be calculated from a knowledge of its orbital period. The relationship becomes very simple if we choose to make our measurements in terms of the astronomical unit (the mean distance between the Sun and the Earth) and the year (the orbital period of the Earth). By way of example, if a planet's orbital period is eight years we can infer that its mean distance from the Sun is 4AU, since $8^2 = 64 = 4^3$.

Kepler's revolutionary ideas were not immediately accepted by supporters of the Copernican theory. Galileo, for one, was still enthralled by circular motions and seems never to have accepted the idea of elliptical motion. However, the system's simplicity and its success in calculating planetary positions ensured that, as the seventeenth century progressed, it became more and more widely accepted throughout the scientific community.

Thus it was that Kepler ended, forever, more than two thousand years of dogmatic adherence to the belief of perfection in the heavens, and the idea that perfect circular motion was the only one possible for celestial orbs. Moreover, the Earth had finally been dethroned from its central position in the scheme of the Universe. Physics, astronomy and philosophy could never be the same again.

3
Universal Gravitation Emerges

Kepler had demonstrated the way in which the planets move, but *why* did the planets move in this way? What motive force could keep the planets moving along elliptical orbits, increasing and decreasing their speeds in accordance with Kepler's second law? Kepler had his own ideas, as we shall see later; but before examining these we must look back to mediaeval times for the first ripples of opposition to the Aristotelian doctrine of force and motion.

One of the first to challenge part of the conventional theory of motion was John Philoponos, a Greek critic of the early sixth century AD. He denied that the resistance of a medium was an essential ingredient in the theory of motion, maintaining that the effect of the medium was merely to slow down bodies. In the absence of a medium, bodies would travel at a finite speed; bodies would not achieve infinite velocity in a vacuum. He also rejected the idea that the speeds of falling bodies were determined by their weights. Furthermore, he opposed the Aristotelian notion that force was transmitted through the air to propel a projectile such as an arrow. Instead, he suggested that a force is impressed on the arrow when it is fired from the bow, and that this force resides in the arrow and keeps it going until the resistance overcomes its motion.

Similar comments were made by later commentators, but the Aristotelian theory continued to hold sway. During the thirteenth and early fourteenth centuries another idea began to emerge, the concept that bodies might possess internal resistance. If real bodies were a mixture of elements, then the proportion of heavy and light elements in a body would determine its motion—whichever element was dominant would determine the direction of motion ("down" for heavy bodies) while the others would tend to counter this motion.

This line of argument led Thomas Bradwardine (*c.*1290–1349), an Oxford scholar, to the interesting suggestion that bodies of similar composition but different weight would fall at the same rate because they contained the same proportion of heavy to light elements, and the ratio of motive force to resistance should be the same.

The idea of an impressed motive force, which had been raised by Philoponos, achieved considerable popularity in the fourteenth century, being developed notably by Jean Buridan (*c.*1300–*c.*1360) of Paris. According to Buridan, when a body is set in motion it is given a certain amount of *impetus* which enables it to continue to move in that direction until the impetus is used up, after which the body should move towards its natural place. Buridan defined impetus as being weight multiplied by velocity, a concept close to our present-day definition of momentum. Although he clearly regarded impetus in terms of a force, nevertheless the concept may be regarded as having paved the way for the important later concept of inertia.

The real upheaval in ideas was initiated by Copernicus, with the publication in 1543 of his heliocentric theory; but, although he dethroned the Earth from the central position essential to the Aristotelian viewpoint, he was unable to produce a new theory of force and motion to account for the observed motions. It was his supporter Galileo who was to lay the foundations of mechanics essential for the establishment of Newton's theory of gravitation.

In addition to his profoundly important astronomical observations Galileo made fundamental advances in the science of mechanics. This work appeared in many of his publications, the most notable of which were *Dialogo . . . sopra i due massime sistemi del mondo, Tolemaico e Copernico* ("Dialogue . . . concerning two world systems, Ptolemaic and Copernican"—henceforth referred to as the *Dialogue*), published in 1632 and banned shortly afterwards; and *Discorsi e dimonstrazioni matematiche intorna à due nuove scienzi* ("Discourses and mathematical demonstrations on two new sciences"—henceforth referred to as *Discourses*) published in Holland in 1638. In both of these works the issues were debated by two central characters, Salviati, representing the modern Copernican viewpoint (and so acting as a mouthpiece for Galileo's own views) and Simplicio, representing the traditional Aristotelian point of view. A third character, Sagredo, took the part of the "unbiased", broad-minded sceptic, who had to be convinced by one view or the other.

The *Dialogue* put Galileo in such bad odour with the authorities of the Catholic Church, especially since Pope Urban VIII thought that Simplicio was a caricature of himself, that he was brought before the Inquisition and made to recant the Copernican heresies which he had been proclaiming. After this event, in June 1633, he was confined effectively to a state of house arrest until his death in 1642.

Among his many achievements, Galileo established the mathematical relationship between the distance travelled by a falling body and the time for which it had been falling. He showed that in the successive seconds after being released the body would fall through 1, 3, 5, 7, ... units of distance, so that the total distance covered at the end of each second would be 1, 4, 9, 16, ... units; i.e., the distance fallen was proportional to the square of the time. This is an example of the simplest form of acceleration, *uniform* acceleration, where equal increments of speed are added each second; the speed of an accelerating body was shown to be proportional to time, rather than to distance as had previously been supposed. The nature of acceleration was one of the central debating issues in mechanics.

He showed, too, that, neglecting the effects of atmospheric resistance, all bodies, regardless of how heavy they are, accelerate towards the ground at the same rate. If a heavy weight and a light weight were to be released at the same instant from the top of a tall tower, they would hit the ground at the same instant. This was contrary to the traditional view that the heavier the object the faster it would fall. (It is said that he tested this hypothesis by dropping weights from the Leaning Tower of Pisa, but this story is almost certainly apochryphal.) Galileo further contradicted Aristotle in showing that a vacuum could exist and in demonstrating how one could in principle be produced. Some thirty years later Robert Boyle tested Galileo's contention about falling weights and showed that in a vacuum jar a feather and a golden guinea fell at equal rates.

In flat contradiction to Aristotelian physics, Galileo showed that a body could undertake two different kinds of motion at the same time. For example, according to Galileo, an arrow fired horizontally from a bow would move forward by equal amounts in equal units of time, and it would also fall towards the ground according to Galileo's law for falling bodies; an arrow fired at an upward angle would follow the same rules. Both of these motions would take place

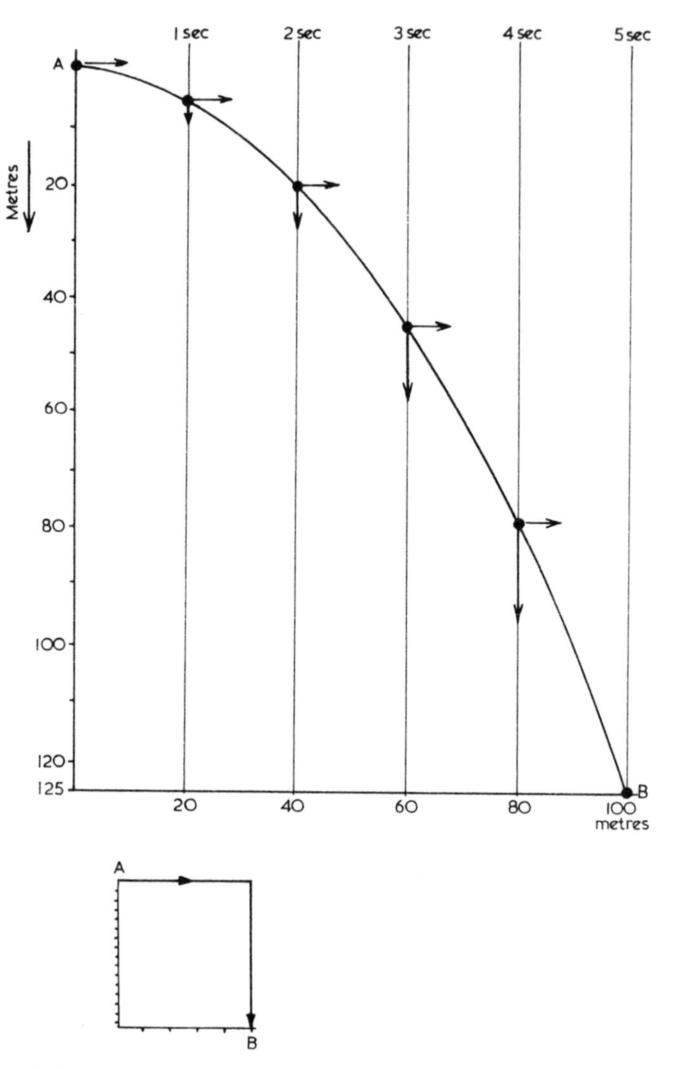

Fig. 9 *The motion of a projectile.* Galileo showed how a body could undertake two different kinds of motion at the same time. If a body were thrown in a horizontal direction from the top of a cliff, it would maintain constant forward velocity thereafter. However, it would also accelerate downwards under the influence of "gravity". Its downward velocity would increase as shown by the vertical arrows. The combination of horizontal and vertical motions lead to the projectile's following a curved (parabolic) path. In the example shown, the forward velocity is 20m/sec, and the height of the cliff 125m; after 5 secs the projectile lands 100m from the base of the cliff (the numbers have been rounded off). Inset is the Aristotelian view of this event: the projectile would first move in a straight line parallel to the ground, and then drop vertically.

at the same time, with the result that the path followed by such a projectile would be a curve known as a parabola (fig. 9). As we have seen, according to Aristotle the projectile would follow one straight line until its forward motion ceased, and then it would drop vertically to the ground.

Galileo maintained that force was *not* necessary to maintain motion, and came very close indeed to discovering what later became known as Newton's first law of motion. In *Discourses* he pointed out that once a velocity was imparted to a body moving on a horizontal plane the body would continue to move in that direction, so long as external causes of acceleration or retardation were removed; i.e., once a body is set in motion it will continue to travel in the same direction unless caused to do otherwise. He argued that, while heavy bodies were inclined to fall and disinclined to rise, they were indifferent to uniform motion on a plane. Once set in motion on a plane a body would be inclined neither to accelerate nor to decelerate. All these points, of course, were contrary to the established opinion that force was necessary to maintain any kind of motion.

What Galileo had arrived at was a form of the law of *inertia* (inertia, in modern terms, being the resistance of a body to any change in its state of motion; i.e., resistance to acceleration or deceleration). But Galileo's concept was of "circular inertia" whereby a body could continue moving on a circular path indefinitely in the absence of forces. On the small scale, on Earth, motion would continue in a straight line but, since the Earth was round, the "flat plane" on which steady motion took place would have to be parallel to the. surface of the Earth. He could not conceive of motion continuing indefinitely in an infinite straight line. The Earth and planets moved around the Sun in circular paths without the action of any force, Galileo argued, because circular motion was natural motion. In the *Dialogue* he argued that a body accelerates towards its goal and is reluctant to leave it: on a circle a body is both approaching and receding from its natural goal, and the net result should be motion at uniform speed. He invoked this notion of "circular inertia" to explain why an object dropped from a tower lands at the foot of the tower rather than reaching the ground some distance to the west: because of its circular inertia the object has a natural tendency to move round at the same rate as the Earth so that, when released, it continues its uniform circular motion around the centre of the Earth while at the same time accelerating towards the Earth's centre.

So, although he did not shake off the fetters of the prevailing doctrine of uniform circular motion of the planets, despite the fact that he knew of Kepler's work, Galileo swept away many of the fundamental pillars of Aristotelian mechanics: he showed that force is not necessary for motion; that bodies can undertake different kinds of motion at the same time; that falling bodies accelerate at the same rate, regardless of their weights; and that a vacuum is possible. Undeniably, he laid the foundations on which Newton was to build the new mechanics.

As to the nature of gravity, Galileo had no real idea; and he took to task those who thought they had. In the *Dialogue*, when Simplicio is asked what causes bodies to fall to Earth, he answers, "... everyone is aware that it is gravity." To which Salviati answers, "You are wrong, Simplicio, what you ought to say is that everyone knows it is called gravity." In other words, Galileo was pointing out that it is all very well to give something a label, but labelling alone does not imply that we understand what that "something" *is*.

Kepler's celestial magnetism

Kepler held fast to the idea that force was necessary to sustain motion, and, therefore, some force must act to push the planets along in their orbits. He maintained that this force must emanate from the Sun, and that it must diminish with distance for, as he argued in the *Epitome* (1621), since the speeds of the planets depend on their distances from the Sun, surely the motive force must originate in the Sun itself. He argued that light itself could not be responsible for pushing the planets along, for if that were so the Earth would stop in its tracks every time an eclipse took place.

He was much influenced by the discovery by the English physician William Gilbert that the entire Earth behaved like a huge magnet, since it was known that the motion of one magnet could cause motion in another, even at a considerable distance. Kepler seized upon and developed this idea, suggesting that magnetic threads emanating from the rotating Sun acted on the planets to push them along, although he was unable to produce a convincing explanation of their elliptical motion. He tended to the idea that there existed some kind of general attraction between bodies, so that distant bodies placed in space would tend to move towards each other and unite. Indeed, he held that space must be a vacuum, and so espoused the possibility that influences could extend across empty space; this was a considerable departure from prevailing opinion. He argued that the ocean tides were due to the influence of the Moon and, to a lesser extent, the Sun.

Although many of his views were confused and incorrect, they were remarkably prescient of Newtonian gravitation in a number of respects; and in this area he proved to be more perceptive than Galileo, who rejected the suggestion that forces emanating from the Sun could move the Earth, and also denied Kepler's suggestion that it was the Moon's attraction which caused the tides in favour of a quite erroneous theory of his own.

Further developments in mechanics

From the time of Kepler and Galileo until Newton's great leap forward the science of mechanics and the concept of gravity were bedevilled by a profusion and confusion of ideas, concepts and misconceptions. Slowly the essential concepts of force, mass, inertia, velocity and acceleration became more clearly defined; at the same time the heliocentric world view became widely established, and the scene was set for a great synthesis of ideas into a coherent whole. Of the many who contributed to the debate and who, in some cases independently, developed important pieces of this elaborate jigsaw puzzle, only a few can be mentioned here.

The great French philosopher René Descartes (1596–1650), in his *Principles of Philosophy* (1644), set out a law of inertia which was essentially the same as what was later to be called Newton's first law of motion. In his view the natural state of motion was uniform straight-line motion: a body would continue to move at uniform speed in a straight line unless something (a force, in modern parlance) acted to change its state of motion. This was a much more complete statement than Galileo's concept of inertia, which was concerned with motion on planes and which, in effect, amounted to a principle of uniform circular motion. According to the mechanical philosophy of Descartes, all changes in motion must be caused by impacts between material bodies. Therefore falling bodies must be pushed back down to earth by streams of subtle (invisible) particles.

A noteworthy attempt to account for the motions of the planets was made by the Italian natural philosopher, Giovanni Alfonso Borelli (1608–1679), who published his ideas in 1666. In essence, he proposed that each planet moves as a result of three causes: (1) a motive force which pushed it along its orbit, (2) an attraction or "appetite", towards the Sun, and (3) a counterbalancing tendency of the planet to recede from the Sun.

He suggested that sunlight was the motive force which pushed the planets along. As the Sun rotated, its light, striking the planets, would exert a weak effect which, over a sufficiently long period of time, would have propelled the planets to their present speeds; the planets would have settled down to a steady motion when their speeds matched the speeds of the impacting particles of light. (It was not clear, on this argument, why the planets moved at different speeds.) Accepting Descartes' hypothesis that bodies continue to move in a straight line unless acted on by some influence, Borelli argued that a force must act to deflect the planets from straight-line paths and keep them in orbit round the Sun. In his view this force must be related to the Sun. If heavy bodies have a natural tendency to fall and unit with the Earth, perhaps, he argued, the planets had a natural "appetite" to unite with the Sun. To prevent the planets falling into the Sun, there had to be some other influence at work—a natural tendency of bodies moving in a circle (or ellipse) to recede from the centre. This was a concept which had been discussed by Descartes. The Dutch physicist Christiaan Huygens (1629–1695) called it centrifugal force.

Like his predecessors, Borelli was not able to make the conceptual leap necessary to see that circular motion is constantly accelerated motion due to the action of one force alone, a force directed towards the centre of the circle. In making this remark we should not be overly critical of Borelli since many of us today still tend to visualize orbital motion in much the same way as he did. We loosely try to "explain" the fact that a satellite remains in orbit around the Earth by saying that the gravitational attraction of the Earth is balanced by an equal and opposite centrifugal force; hence the satellite "stays up there". Although as a classroom exercise we can obtain correct formulae to describe circular motion by these means, in fact this is a wholly erroneous viewpoint. Centrifugal force is merely an *apparent* force; the only *real* force acting on the satellite is the force of gravity. Consider the familiar analogy of whirling a stone round on the end of a piece of string. In order to keep the stone moving in a circle we have to exert a force—known as centripetal force (the "centre-seeking" force)—on the string; i.e., we have to maintain a pull on the string. If we cut the string, the stone does not fly radially outward (as it ought to do if a force directed away from the centre were acting), but flies off along a tangent; that is, it continues in a straight line at uniform speed in the direction in which it was heading at the instant the string was cut, so obeying the law of inertia as set out by Descartes.

Seventeenth-century natural philosophers were hampered in their efforts by confusion over the definitions of fundamental quantities. Velocity was well understood, the concept of acceleration had been clarified by Galileo, but the quantities which we, today, call mass, weight, force and inertia were poorly formulated. Considerable clarification seems to have been achieved by the French philosopher Edmé Mariotte (1620–1684): mass, the quantity of matter in a body, depended on that body's size and density, and was a different thing from the heaviness of the body. Moreover, he argued, each body had a resistance to the acquisition of motion which depended on its mass. If two balls were placed on a horizontal plane and struck with the same blow, the less massive of the two would acquire the higher speed. The concept of "heaviness", which was associated with the fall of bodies towards the Earth, would not apply here, and so there had to be some other quality. He dismissed the possibility that air resistance had any significant effect since then it would be the *size* of the ball which governed the resistance: although a two-pound ball of lead is smaller than a one-pound ball of wood, it has a greater resistance to acceleration. We see here the emergence of the concept that, although weight and mass are related (a heavier body has more mass than a lighter body), it is mass which determines the resistance of a body to motion. The concepts of mass and inertia (resistance to acceleration) as discussed here were essentially the same as those used in Newton's mechanics.

The debate on the nature of gravity (the term was in widespread use in the seventeenth century) continued. Was gravity an external agency acting on bodies, or was gravitation an internal property possessed by bodies? Was it an attractive influence which could act between widely separated bodies, or was it due to the impact of invisible particles? To the likes of Huygens, and other followers of Descartes, the idea of attraction—the action of one body on another at a distance—was anathema. The very idea reeked of occultism, of mystical influences. Surely, only the action of particles on bodies could affect their motion.

Pierre Gassendi (1592–1655) was one who regarded gravitation as an external force which acted on bodies to draw them to the Earth. He considered both gravity and magnetism to be due to streams or "threads" of imperceptible particles which emerged from the Earth and dragged bodies back towards their source. In a completely empty space, where no gravitational threads were present, bodies once set in motion would continue to move at uniform speed in a straight line; but near to a body like the Earth bodies were acted upon by gravitational threads and dragged from their linear trajectories. Gassendi's ideas did not seem to have any direct effect on the development of Newtonian gravitation, but they are mentioned here to illustrate that at this time many ideas concerning gravity and celestial motion were current, and that various individuals separately devised, or closely approached, concepts which formed parts of Newton's great synthesis.

Newton's theories emerge

Three of the leading scientific personalities of the mid-seventeenth century in England were Robert Hooke (1635–1703), the architect Christopher Wren (1632–1723), and Edmund Halley (1656–1742), later to become Astronomer Royal and to achieve lasting fame for his work on comets. They were prominent members of the Royal Society, officially founded in 1660, where learned men met to discuss scientific problems and to plan experiments. Among the problems discussed at length by Wren, Halley and Hooke was the question of what law of attraction would cause the planets to follow elliptical orbits around the Sun. By 1684, Halley was firmly of the opinion that the attractive force diminished with the square of distance (i.e., if the distance were doubled, the force would be reduced to one quarter of its original value). This seemed to be a reasonable supposition, bearing in mind that it was well known that light behaved in this way; moreover, if any influence spreads out symmetrically in all directions from a source, the area over which it spreads increases with the square of distance and so, most probably, its strength would decline in proportion to the area over which it was spread. Halley and the others were unable, however, to prove this assertion by a mathematical demonstration that, from such a law of attraction, elliptical orbits would result.

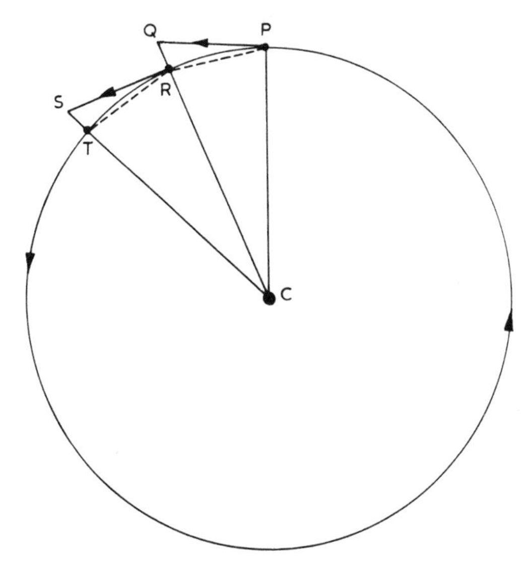

Fig. 10 *Circular motion.* Imagine a body following a circular orbit under the influence of an attractive force directed towards the centre of the circle (a "central force"). At point P the body will be moving in direction PQ and, if the attractive force were removed, it would continue to move in that direction at constant velocity in accordance with Newton's first law. In the time which the body would have taken to reach point Q, it has "fallen" under the action of the central force a distance QR to the point R on the circle. At point R the body will be moving in the direction RS, but the effect of the attractive force will be to bring it instead to point T. Thus a body following a circular path is continually falling towards the centre but never getting any closer to it. Newton applied this line of reasoning to the motion of the Moon.

In August of that year, Halley went to Cambridge to consult Isaac Newton (1642–1727), the Lucasian Professor of Mathematics. When he asked Newton what would be the path followed by a planet acted on by an inverse-square force of attraction towards the Sun, Newton immediately replied that it would be an ellipse, and that he had proved this some time previously. Newton had mislaid his calculations, but promised to repeat them and send them on to Halley, which he did, a few months later. This encounter reawakened

Newton's interest in gravitation and planetary motion. With constant encouragement from Halley, Newton was persuaded to gather together his work into the treatise, *Philosophiae Naturalis Principia Mathematica* which, published in 1687, ranks as one of the most fundamental and influential books ever written.

Newton's investigation of gravity really began in 1665. He had returned to his home in Woolsthorpe, near Grantham in Lincolnshire, because Cambridge University had been closed on account of the Great Plague. It is said—and it may even be true—that by observing the fall of an apple from a tree in the garden he was led to consider whether the force which caused the apple to fall might not be the same force as held the Moon in its orbit round the Earth. He proceeded to investigate the matter.

Newton deduced (as did Huygens, independently, some years later) that a body moving in a circular path must be subject to a constant acceleration due to a constant force acting towards the centre of the circle (fig. 10). A body moving on a circle was continually "falling" towards the centre but, because of its "sideways" velocity, never getting closer to or further from that centre. From Kepler's third law of planetary motion it was known how the orbital period and distance of a planet from the Sun were related. Combining this information with his formula for circular motion, Newton found that the force responsible for keeping the planets in their orbits (assuming circular orbits) was inversely proportional to their distances from the Sun. He decided to find out whether the forces which controlled the fall of the Moon and the fall of the apple were one and the same by comparing the acceleration of the Moon in its orbit and the apple in its fall, assuming that the force responsible was an inverse-square force.

He inferred that the important distance to be considered was from the Earth's centre rather than from its surface, although he was not able to prove this assertion until much later. Knowing that the distance of the Moon is about sixty times the radius of the Earth, and knowing the period of time taken by the Moon to travel round the Earth, it was a straightforward matter to calculate the Earthward acceleration of the Moon; i.e., the distance through which it "falls" every second, or every minute. The acceleration of a falling body close to the surface of the Earth was known quite well experimentally. If the Moon were sixty times further from the centre of the Earth than was the apple, the apple should be subject to an acceleration due to gravity of 60×60 (3600) times greater than that experienced by the Moon. In one second, therefore, the apple would fall through a distance equal to that through which the Moon would fall in one minute.*

Newton made the comparison and found that the two rates "answered pretty nearly". In fact, Newton's original calculation suggested that the stone's acceleration was over 4,000 times that of the Moon, but this discrepancy turned out to be due mainly to the inaccurate value of the Earth's radius available to him at that time. When he repeated the calculation in 1684 he used improved values, and when the results were published in the *Principia* the agreement was within one per cent.

This suggested strongly to Newton that the *same* force which controlled the motion of the planets around the Sun governed the motion of the Moon and the falling of objects to the ground—the force which was known commonly as gravity.

Before he could develop the theory further he had to develop the necessary mathematical tool, that branch of mathematics known as calculus. (The calculus was developed independently, and in a more convenient form, by the German mathematician Gottfried Leibniz (1646–1716), and there was some international wrangling as to who should have credit for inventing it.) By 1679 Newton had demonstrated that elliptical orbits would result from motion subject to an inverse-square force directed towards one focus of the ellipse, but it was not until 1685, with the aid of his calculus, that he was able to prove that it was permissible to treat the gravitational attraction of the Earth as if all of its mass were concentrated at a point in its centre. When this was shown, he had at last justified the way in which he had made his comparison between the accelerations of the Moon and the apple.

His *Principia* appeared in three parts: Book I set out the general principles of Newton's system of mechanics; Book II dealt with other aspects of physics; and Book III, "The System of the World", dealt with the application of his laws to the Universe. Book I

*Although the acceleration of the apple is 3,600 times greater than that experienced by the Moon, the law of free fall established by Galileo showed that the distance travelled was proportional to the degree of acceleration and the time *squared.*

contained his three laws of motion, laws which are still of fundamental importance today. In modern parlance, they are as follows:

First law: Every body continues in its state of rest or of uniform motion in a straight line unless acted upon by a force;

Second law: The change in motion is proportional to the force and takes place in the direction in which the force is acting; i.e., the acceleration a produced in a body of mass m by a force F is related by the expression $F=m \times a$ (force = mass × acceleration).

Third law: To every action there is an equal and opposite reaction (for example, if you push against a wall, the wall pushes back against you with an equal and opposite force).

The first and second laws finally laid to rest the Aristotelian doctrine of force and motion. In the clearest possible terms, Newton stated that force was not required to maintain motion, that the natural state of bodies was to be at rest or in uniform straight-line motion, and that force was required only to *change* a state of motion. Newton's work established the concepts of force, mass and inertia (resistance to acceleration) in the way in which we use these terms today. He paid considerable tribute to Galileo as the "father" of the first two laws, but his tribute was too generous by far. Although Galileo came close to the law of inertia, Newton's first law, he did not accept the idea of indefinite motion in a straight line, and, although he laid the foundations for the second law with his work on the acceleration of falling bodies, by no means did he arrive at anything resembling a statement of that law.

In Book III Newton demonstrated that the motion of a body following a conic (circular, elliptical, parabolic or hyperbolic) orbit required the action of an inverse-square force directed towards a fixed point, and showed, conversely, that a body moving under the action of an inverse-square force would obey Kepler's laws. He pointed out that the same force accounted for the motion of the Moon, the planets, the acceleration of falling bodies, the motions of the satellites of Jupiter, and the ocean tides.

Since on Earth all bodies fall with equal acceleration, regardless of their masses, from Newton's second law it was clear that for this to be so the force acting on the falling body must be proportional to the mass of that body; i.e., the more massive the body, the greater

Sir Isaac Newton (1642–1727), the English mathematician and physicist who, it has been argued, was the greatest scientific figure in history. Best known for his formulation of the laws of motion and of universal gravitation, he made valuable contributions to many branches of science, including optics. (BBC Hulton Picture Library.)

Albert Einstein (1879–1955), whose Special and General theories of Relativity changed our entire conception of the nature of space, time, and gravitation. (BBC Hulton Picture Library.)

the force required to accelerate it by a given amount ($a=F/m$, and so for a to be constant for all masses, then F must be proportional to m; a 2 tonne mass requires twice as much force as a 1-tonne mass to attain the same acceleration). By the third law it followed that if one mass, m, was acted upon by another mass, M, then an equal and opposite force of attraction would be exerted by m on M (for example, if the Earth attracts the Moon, so must the Moon attract the Earth). Consequently the mutual force of attraction must depend on the masses of both bodies.

Thus gravitation was established to be a force of attraction proportional to the masses of the attracting bodies, and inversely proportional to the square of the distance between them. For masses m and M separated by distance r the force of gravity is given by the expression $F=GmM/r^2$, where G is a constant of proportionality known as the *universal gravitational constant*. The value of G determines the strength of the force of gravity. It is one of a number of fundamental constants of nature, numbers whose values determine the behaviour of the Universe and all that it contains. Whether or not G is constant *for all time* is a matter which is the subject of much debate, as we shall see in Chapter 12.

The term "mass" as it appears in Newton's second law is the *inertial mass*, a measure of the resistance of a body to any change in its state of motion. If the same force is applied to two bodies of different masses, then the less massive will acquire a greater acceleration than the more massive (pushing a wheelbarrow is easier than pushing a 'bus). However, the term "mass" as it appears in Newton's law of universal gravitation is the "gravitational mass", a measure of what one might call the "quantity of gravity" possessed by a body. There is no obvious reason why the two "masses" should be identical. After all, we can regard gravitational mass as the gravitational equivalent of electrical charge; two bodies of the same inertial mass can have very different electrical charges, and will accelerate by different amounts under the influence of an electrical field. That all bodies in the Earth's gravitational field fall with the same acceleration is possible only if the ratio of gravitational mass (which is involved in determining the accelerating force) to inertial mass (which determines the resistance to acceleration) is precisely the same for them all. Newton carried out experiments to see if there was any difference in this ratio for different kinds of bodies. He found none, nor has any difference been found since—even in experiments capable of distinguishing a difference of one part in a million million. Since the two masses are always in the same proportion, units of measurement have been chosen to make them equal. The fact that inertial and gravitational masses are equal is known as the *principle of equivalence.* As we shall see in Chapter 5, this principle forms a key foundation of the General Theory of Relativity.

Successes of Newton's theory

Apart from its success in dealing with falling bodies and planetary motions, Newton's theory had great successes in other directions, too.

The ocean tides were accounted for, in general terms, as being due to the *difference* in the attraction exerted by the Moon on the solid globe of the Earth and the ocean masses on opposite sides of our planet, the net result being to build up humps of water on the sides facing towards and away from the Moon.

Newton predicted that, because of its rotation, the Earth should bulge slightly at the equator and be flattened at the poles, and showed how this departure of the Earth from being a perfect sphere accounted for the phenomenon of precession, whose effects were first discovered nearly 2,000 years earlier by the Greek astronomer Hipparchus. Precession is the phenomenon whereby the axis of the Earth slowly shifts in position in such a way that the celestial pole traces out a circle on the sky. If the Earth were a perfect sphere this would not happen, but the gravitational attraction of the Sun and the Moon on the equatorial bulge causes the Earth's axis to move round just as a spinning top's does when tilted from the vertical.

Edmund Halley, Newton's great supporter, studied the appearances of the comets of the years 1456, 1531, 1607 and 1682, and came to the conclusion that they shared the same elliptical path, the slight discrepancy in the intervals between successive returns being due to the disturbing influences of the planets on the comet's path. He predicted, using Newton's laws, that the comet would return in 1758. When this prediction was fulfilled the comet was named in his honour, although he himself was long dead.

The difficult problems of the perturbations of the planets due to their mutual gravitational attractions was tackled, notably by French and German mathematicians, and by the end of the eighteenth century the science of celestial mechanics, based on Newton's laws of motion and gravitation, had successfully encompassed many complex problems relating to motion in the Solar System. A major triumph for celestial mechanics would come with the successful prediction of the position of a new planet beyond the orbit of Uranus. The calculations, carried out independently in England by J. C. Adams (1819–1892) and in France by U. Leverrier (1811–1877), were based on perturbations of the orbit of Uranus by the hitherto undetected planet. Neptune, as it came to be known, was discovered in 1846 by Galle and d'Arrest of the Berlin Observatory.

Observations published in 1803 by William Herschel (1738–1822), who in 1781 had discovered Uranus, showed that a number of stars, including Castor (one of the two bright stars in the constellation of Gemini, the twins) consisted of two stars which slowly revolved around each other under their mutual attraction: such pairs of stars are called binaries (see Chapter 6). Subsequent observations have shown that the stars making up binary pairs move according to Kepler's laws and the predictions of Newtonian gravitation. Before the nineteenth century was halfway through, Newton's theory of gravity had been shown to operate throughout the known Universe.

Newtonian gravitation was truly a *universal* force. With the acceptance of Newtonian theory came the end of those deep-rooted ideas of Greek and mediaeval times that there was one law for the heavens and another for objects here on Earth. The door was opened for the development of a scientific approach to understanding the Universe and all its constituent parts in terms of a few basic laws and forces which operate in the same way here on Earth, in the laboratory, as they do throughout the cosmos.

Only one aspect of the theory seemed to be unsatisfactory: Newtonian gravitation was a force of attraction which acted across vast distances in the Universe. However, the *nature* of this force remained a mystery, and it was on these grounds that the theory was initially criticized, particularly in continental Europe where the tradition of mechanistic philosophers such as Descartes was strong—mysterious forces which acted across empty space had no place in such a philosophy. Newton himself steadfastly refused to speculate on the actual nature of the gravitational force. The fact remained that his theory *worked*, and it soon came to be widely accepted, even by those who were strongly critical of the concept of a force which could "act at a distance" across empty space.

Well over two centuries were to pass before the emergence of any alternative theory capable of challenging and superseding Newtonian gravitation. The new theory was Relativity, the development of which is discussed in the next two chapters. It is a theory which is superior to Newton's only in certain extreme circumstances, a theory which modifies and extends rather than totally replaces Newtonian theory. Moreover, the measure of Newton's great achievement is that even today Newtonian mechanics is still more than adequate to deal with all but the most esoteric problems. If you wish to send a man to the Moon, or a spaceprobe to the outermost planets, the theories of force, motion and gravitation published by that remarkable scientific genius in 1687 are more than adequate to deal with the problem.

4
Upheaval

The laws of Newtonian mechanics hold true for all observers who are in uniform relative motion. What does this phrase mean? In conducting experiments and making observations of events we require a frame of reference, a standard of position and time measurement against which to judge the position and time of each particular event. We usually think of space as having three dimensions, and of objects in space as having "length", "breadth" and "height", corresponding to the three directions in space of "along", "across" and "up". In tales of hidden treasure, the ill gotten gains of the pirate crew may be found with the aid of a fortuitously acquired map which advises the bearer to start at a particular point, march X paces north, Y paces east, and then climb to a height Z. In this way, the spatial coordinates of the treasure (or of any point) may be described (fig. 11).

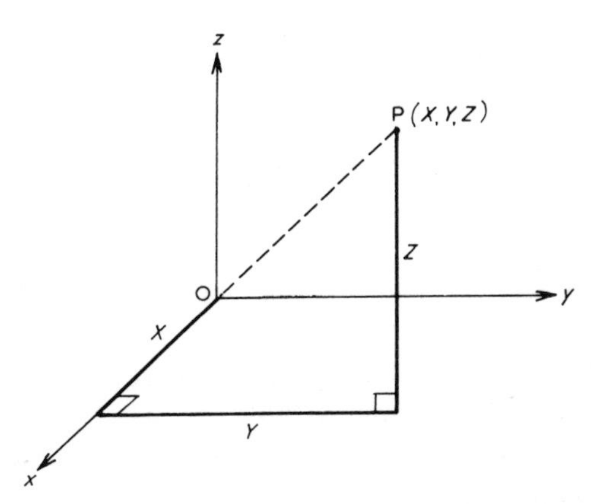

Fig. 11 *Coordinates in space.* The position of the point P in relation to an observer at O is specified by reference to three mutually perpendicular axes defining the x-direction ("along"), y-direction ("across") and z-direction ("up").

The time of an event may be determined by reference to a clock, and it was a fundamental aspect of Newton's philosophy that there existed *absolute time* in the Universe. In his own words, written in the *Principia*, "Absolute, true, and mathematical time, of itself, and from its own nature, flows equably without relation to anything external ..." In other words, time flowed past at a uniform rate, passing on from one instant of absolute time to the next, regardless of what was happening in the Universe. For an observer, time would carry on in its smooth, relentless flow, regardless of where he was located or how fast he was moving. This is still the "common sense" view of time today; an hour is still an hour whether you are sitting in an armchair or flying the Atlantic in a supersonic aircraft.

A frame of reference, then, can be regarded as a rigidly defined means of measuring position and time (e.g., three metre rules set at mutual right angles, for measuring position, and a clock, for measuring time). An *inertial* frame of reference is one free from acceleration or rotation. An observer located in such a frame of reference—i.e., an observer who is stationary or moving at uniform speed in a straight line—is called an inertial observer. It is easy to see that two observers in uniform *relative* motion will find that the laws of Newtonian mechanics hold true in both of their frames of reference. For example, consider one observer standing beside a railway track while the other is located inside a high-speed train travelling at uniform velocity. If the observer on the train throws a ball along the corridor at uniform speed, the stationary observer will also see that ball moving with uniform speed; admittedly the speed he will measure will be greater (being the speed of the ball plus the speed of the train) but the ball will be travelling in accordance with Newton's first law in both frames of reference. Likewise, if the moving observer drops the ball he will measure the same rate of acceleration (due to the gravitational force exerted by the Earth) as would the stationary observer if *he* were to drop a ball. Thus the second law (force = mass × acceleration) is also obeyed in both frames. Likewise, the third law will be the same for both observers.

The problem with all this is twofold: what is so special about inertial frames of reference?; and against what can one judge whether or not an individual frame of reference is truly inertial? An inertial frame is one which is at rest or in uniform straight-line motion (in Newton's view), but at rest or in uniform motion with respect to what? Newton's answer was "Absolute space". In his own words, "Absolute space, in its own nature, without relation to anything external, remains always similar and immovable." Absolute space was a fundamental background to the Universe, an absolute standard of rest relative to which, in principle if not in practice, it should be possible to determine the absolute motion of a body, from one "absolute place" to another. Absolute space was indifferent to uniform straight-line motion, but acted to resist the acceleration of bodies. The inertia of bodies, in Newton's view, arose because of the action of absolute space.

Objections to the concepts of absolute space and absolute time were raised by contemporaries and near-contemporaries of Newton, such as Leibniz and Bishop Berkeley on the grounds that they ascribed to space and time a physical existence in themselves. The alternative view, championed by Leibniz, was the *relational* theory of space and time, whereby "space" was simply the separation between bodies and "time" was merely the succession of events; according to this view neither space nor time had an independent existence in their own right and relative motions were all that mattered.

Absolute space had the property that it acted on bodies (to resist their acceleration) but was itself unaffected by matter. Of this, Einstein later remarked, "It conflicts with one's scientific understanding to conceive of a thing which acts, but cannot be acted upon." Newton attempted to identify absolute space—the absolute standard of rest—with the centre of mass of the Solar System; later commentators associated it with the frame defined by the "fixed stars"; and, in modern terms, the logical identification would be with the frame of reference in which the distant galaxies are seen to be receding uniformly away from each other.

Ernst Mach (1838–1916), in 1872, argued that the property of inertia had nothing to do with absolute space, as such, but arose from some kind of (unspecified) interaction between each individual body and all the other masses in the Universe. If there were no other masses, Mach argued, an isolated body would have no inertia; this contrasted with Newton's opinion that the body would still have inertia because of the effect of absolute space. The opinions expressed by Mach were to greatly impress Einstein, who coined the term *Mach's principle* to describe the hypothesis.

During the nineteenth century great progress was achieved in the study of electricity and magnetism. Electrical phenomena such as sparks, lightning, and the behaviour of the Leyden jar (which stored electricity) were at first regarded as being different from magnetic phenomena, which were associated with the behaviour of certain kinds of rocks, compass needles and the like. However, the Danish natural philosopher Hans Christian Oersted (1777–1851) and the French physicist André Marie Ampère (1775–1836) showed by experiment that an electric current moving in a wire gave rise to a magnetic effect which deflected compass needles. A few years later, in 1831, the great English experimenter Michael Faraday (1791–1867) showed, conversely, that a moving magnet induced an electric current in a wire.

Electricity and magnetism were successfully married together into the electromagnetic theory of the outstanding Scottish mathematical physicist James Clerk Maxwell (1831–1879). Following an earlier suggestion by Faraday, he developed the concept of the *field.* Each charged particle was considered to be surrounded by a field, an invisible aura which acted upon other charged particles placed within it; i.e., the field of one charged particle exerted a force upon another charged particle. This idea was strikingly different from the Newtonian concept of gravitation, which was seen as a force which acted directly across a distance between one mass and another. In Maxwell's theory it was the strength of the field at a given point which affected the motion of a particle placed at that point.

Maxwell's equations describing the electromagnetic field predicted that the motion of charged particles should generate waves —electromagnetic waves—which would travel through space with a speed which turned out to be equal to the speed of light, 300,000km per second. Furthermore, these waves could have any value of wavelength (the distance between successive wave crests;

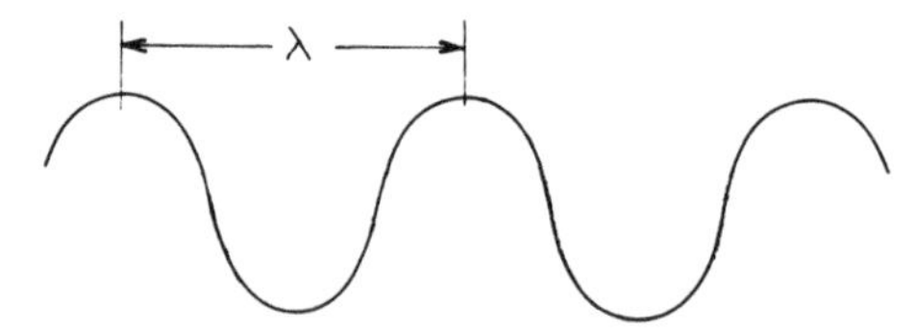

Fig. 12 *A light wave*. A light wave is visualized as being like a wave on water. The distance between two crests defines the wavelength, conventionally denoted by the symbol λ; the number of crests per second is the frequency.

see fig. 12): while light waves were one form of electromagnetic waves, clearly waves of longer or shorter wavelength could exist. Heinrich Hertz (1857–1894) succeeded in 1888 in transmitting and receiving waves of much longer wavelength—radio waves. Today we are familiar with wavelengths ranging from less than one million millionth of a metre to many kilometres, and we have divided up this *electromagnetic spectrum* (fig. 13) into arbitrary bands known as gamma-rays, X-rays, ultraviolet, visible, infrared, microwave and radio radiation. All these radiations travel through space at the speed of light (about 300,000km per second) and they are all of the same nature.

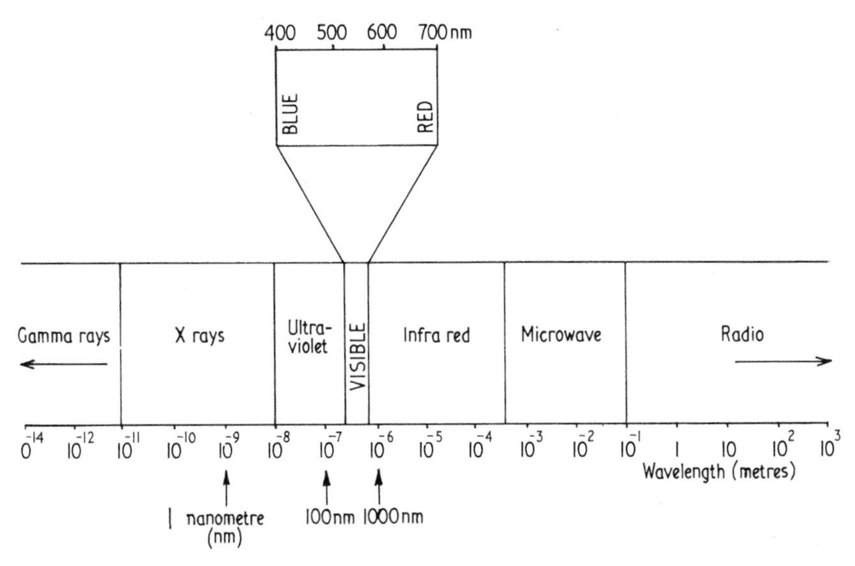

Fig. 13 *The electromagnetic spectrum.* The complete range of possible wavelengths associated with electromagnetic waves, divided somewhat arbitrarily into a number of bands, ranging from gamma rays, of shortest wavelength, to radio waves, of longest wavelength. Visible light spans a narrow range of wavelengths from about 400 nanometres (nm; 1nm is one billionth of a metre) to about 700nm. The different colours correspond to different wavelengths.

Sound waves travel in air, water waves in water. It is difficult to conceive a wave without some kind of medium within which it propagates. Accordingly, Maxwell resurrected the old idea of an aether, filling all space, which served as the carrier of these waves. The frame of reference defined by the stationary aether came to be regarded as the absolute standard of rest, and became identified in the minds of physicists with Newton's absolute space.

Experiments were then devised to attempt to measure the speed of the Earth relative to the aether. The most famous of these experiments was the Michelson–Morley experiment, first performed in 1881 by the US physicist Albert Michelson (1852–1931), and in 1887 with improved apparatus by he and Edward Morley (1838–1923). The principle of the experiment is illustrated by the following analogy. Imagine a race between two power boats, each capable of precisely the same speed, on a river flowing at a uniform rate. Boat A has to cross the river to the far bank and then return to the starting position, while boat B has to travel an equal distance downstream and then back against the current. Which boat will win the race? As is explained in fig. 14, boat A will win every time.

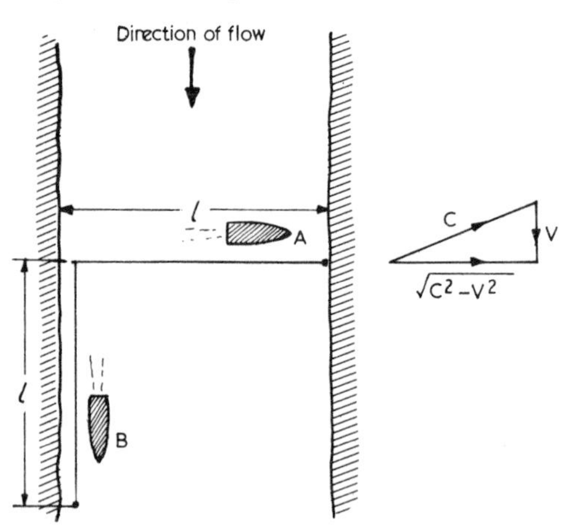

Fig. 14 *A hypothetical powerboat race.* The river flows down the page with velocity v, and both powerboats can maintain a steady speed c, which is greater than v. The width of the river is l. A has to sail directly across the river and back to the start, while B has to sail downstream a distance l and then back up to the start. As shown on the right, A has to aim slightly upstream to allow for the current and will travel directly across the river at a rate rather less than c; using the theorem of Pythagoras, this velocity will be $\sqrt{c^2-v^2}$. A's time to complete the course is thus $2l/\sqrt{c^2-v^2}$.

On the downstream leg B makes good time, travelling at a speed of $c+v$, since the current is with him; but against the current on the way back he will achieve a speed of only $c-v$. B's travel time will be $l/(c+v)+l/(c-v)$ which, with a little elementary algebra, becomes $2lc/(c^2-v^2)$.

Dividing A's time by B's time we arrive at $\sqrt{1-v^2/c^2}$, which must *always* be less than 1, unless $v=0$ (i.e., unless the stream has no current). Therefore A *always* wins the race.

In the discussion in the text of the Michelson-Morley experiment, c denotes the velocity of light and v that of the aether relative to the Earth.

The argument regarding the aether went as follows. If the Earth is moving through the aether, then from our point of view the aether should seem to be flowing past, like the stream in the analogy. If light moves through the aether at a constant velocity (as Maxwell's equations suggested), a ray of light sent in the direction of the Earth's motion, then reflected back to its starting point, should arrive later than a similarly treated ray sent across the same distance but at right angles to the Earth's motion (like boats B and A). The experimental apparatus is illustrated in fig. 15. Since the Earth moves around the Sun at a speed of about 30km per second, then sometimes during the orbit the aether must be flowing past the experimental apparatus at similar speed; the sensitivity of the apparatus was such that motion through the aether of a few kilometres per second was well within the range of measurement. In fact, no matter how often the experiment was repeated, *no difference whatever* was found in the arrival times of the light beams. In terms of the

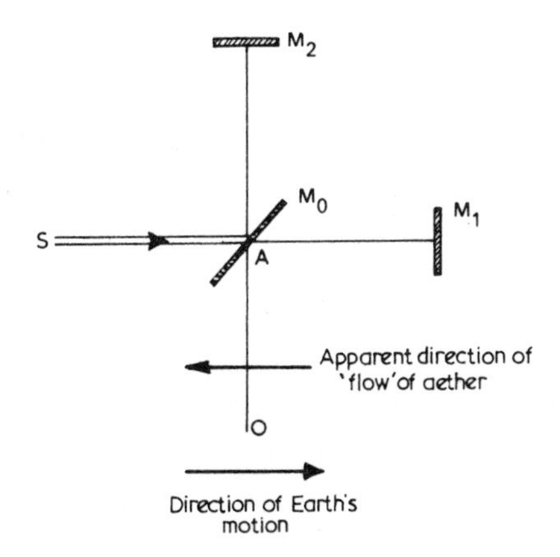

Fig. 15 *The Michelson–Morley experiment.* This simplified diagram illustrates the basic principle of the experiment. A beam of light travels from source S to a half-silvered mirror M_0, which allows half the beam to travel straight on to mirror M_1 while reflecting the other half to mirror M_2, both mirrors being at the same distance from A. The reflected beams from M_1 and M_2 again meet M_0, and part of each travels to the observer O. If light were to move at a constant speed relative to the aether, and if the Earth were moving through the aether as shown, then the aether would appear to flow past the apparatus in the direction of the heavy arrow (rather like the flow of the stream in fig. 14). That being the case, the beam from M_1 should take longer than the other beam to complete its journey, and the observer at O should be able to measure the extent to which the beams have got out of step. All attempts to make this measurement failed, indicating that the postulated "flow" of the aether had no measurable effect on the two light beams.

powerboat analogy, the experimental result seems to make no sense at all, for it implies that the speed of the river makes no difference whatever to the time taken for the two boats to complete their journeys.

Gradually it became apparent to physicists that *no* experiment was capable of showing the Earth's motion through the aether. When this conclusion had reluctantly been accepted attempts were made to preserve the notion of the all-pervading aether by means which were somewhat devious, to say the least. During the last decade of the nineteenth century the Dutch physicist Hendrik Lorentz (1853–1928) and the Irish physicist George Fitzgerald (1851–1901) independently suggested that movement through the aether would shrink measuring-rods and slow down clocks by just the right amount to prevent that motion from being determined; if the distance up and down stream were, in fact, shorter than the distance across the stream the null result would be "explained"; however, since any device used to check the former length would be reduced in the same proportion, the change in distance could not be measured. While this was a very convenient way of getting round the fact that motion through the aether could not be observed it also implied, as the French mathematician Jules Henri Poincaré (1854–1912) pointed out, that the aether, even if it existed, could never be detected. There seemed to be no way, even in principle, of deciding whether or not the aether existed, and so it seemed sensible to consider only *relative* motions, rather than absolute ones.

There was another problem. Whereas the laws of mechanics took the same form and were equally valid in all inertial frames of reference, whatever their relative motion, this did not seem to be the case for Maxwell's electrodynamics. The speed of light through the aether was a central element in Maxwell's equations, and so these equations were based upon one particular frame of reference, the frame of the aether. Observers moving relative to that frame would see different effects, and the laws of electrodynamics would have the same form only for those observers who were at rest relative to the aether. Why should one branch of physics (mechanics) be indifferent to uniform relative motion while another (electrodynamics) was dependent on the speed of the observer?

The Special Theory of Relativity

In 1905 Albert Einstein (1879–1955), then an obscure patent clerk in the Swiss Patent Office in Berne, published his Special Theory of Relativity, a theory which resolved the problems of the Michelson–Morley experiment and the theory of electrodynamics, and which at the same time swept away the creaking foundations of the classical concepts of space and time. The theory rested on two fundamental postulates.

The first was the *relativity principle*, that all inertial frames are fully equivalent for the performance of all physical experiments. This implies that provided a laboratory is neither accelerating nor rotating—i.e., if it is moving at uniform velocity—its motion will have no effect whatever on the results of experiments carried out inside it. All observers in uniform relative motion will deduce the same physical laws from the results of their experiments. The relativity principle abolished the distinction between the behaviour of the laws of mechanics and the laws of electrodynamics, and made the whole idea of an aether and of Newton's absolute space utterly redundant. The basis of more than two centuries of established physics was swept away at a stroke.

The second postulate was that the speed of light was constant in all inertial frames of reference. In other words, the speed of light measured by any observer in uniform motion is unaffected by the relative velocity of the source and the observer. This is totally contrary to what common sense would decree. If two trains are approaching each other, each moving at 100km per hour, we would have no hesitation in saying that the relative velocity of the two trains is $100+100=200$km per hour. If a spacecraft travelling at 100,000km per second is approaching a source of light, and if light is travelling from that source at a speed of 300,000km per second, common sense would suggest that the relative velocity of the light and the spacecraft, measured by an observer on the spacecraft, would be 400,000km per second. According to Special Relativity, however, the observer would find that the speed of the incoming light was precisely 300,000km per second. The speed of the spacecraft would have no effect whatever on the measured velocity of that ray of light.

Although these conclusions may appear to be absurd, they are entirely in accord with the null result of the Michelson–Morley experiment, and the failure of all other experiments designed to show that the motion of the source or of the observer has any effect on the measured velocity of light. The Universe is constructed in such a way that all observers will measure the same value of the speed of light. In the face of the evidence resulting from many carefully constructed experiments we have no option but to accept these conclusions, however much they go against that limited body of local experience which we call "common sense".

The observations of position and time made by observers in uniform relative motion are related by a set of equations known as the *Lorentz transformations.* Einstein adapted the equations of Newtonian mechanics so that they, Maxwell's equations, and the speed of light were invariant (i.e., took the same form in different coordinate systems) when related *via* the Lorentz transformation. Armed with these equations, physicists can work out how the observations made by different observers relate to each other.

A number of interesting consequences follow from accepting the postulates of Einstein's theory.

Length contraction: As Lorentz and Fitzgerald had hinted earlier, the length of a moving object is affected by its motion. If a spacecraft flies at very high velocity past an observer (whom we shall describe as "stationary" although accepting that nothing in the Universe can claim truly to be "fixed") it will appear to the stationary observer to have shrunk in length by an amount which depends on the velocity of the spacecraft. The closer the spacecraft approaches the velocity of light the more pronounced this effect becomes until—if it were possible for it to move precisely at c, the velocity of light—it would be observed to have zero length. Values of length contraction for different velocities are given in Table 1.

The inhabitants of the spacecraft will be unaware of this effect. To them, everything is normal, for the "contraction" will have affected everything in the spacecraft in equal proportion. They might notice, however, that the observer's spacecraft has apparently shrunk in length—for his speed relative to them is of course the same as their speed relative to him.

Table 1 Length contraction, time dilation and mass increase

	Length contraction	Mass increase	Time dilation
Velocity (v) as a fraction of the speed of light (c): v/c	Length (l) of a moving object as a fraction of its rest-length (l_0)	Mass (m) of a moving object compared to its rest-mass (m_0)	Length of a time interval (Δt) measured on a moving clock compared to the interval (Δt_0) measured on a "stationary" observer's clock
0	1.000	1.000	1.000
0.1	0.995	1.005	0.995
0.5	0.867	1.155	0.867
0.7	0.714	1.400	0.714
0.9	0.436	2.294	0.436
0.99	0.141	7.089	0.141
0.999	0.045	22.366	0.045

The terms rest-length (l_0) and rest-mass (m_0) refer to the length and mass which a moving object of observed length l and mass m would have if it were stationary relative to the observer. The relationships between l and l_0, m and m_0, and Δt and Δt_0 are as follows:

$$l = l_0\sqrt{1-v^2/c^2};$$

$$m = m_0/\sqrt{1-v^2/c^2};$$

$$\Delta t = \Delta t_0\sqrt{1-v^2/c^2}.$$

Time dilation: The rate at which time passes on a fast-moving spacecraft is slower than the rate of time's passage as measured by a "stationary" observer. If an Earth-based observer could see the clocks on a fast-moving spacecraft he would come to the conclusion that they were running slow compared to his own clock. Again, this effect becomes rapidly more obvious as velocities approach that of light (Table 1) until, if it were possible for a spacecraft to move at c, time inside it would be seen to stand still compared to the stationary observer's time. Time measured on a clock which an observer carries with him is known as *proper time*; all other clocks in motion relative to that observer run slower than his proper clock.

This affects everything on board—including atomic processes and, above all, the biological clocks of the crew. If this were not so the relativity principle would be violated, for it would be possible for the crew to make measurements which would reveal their state of motion; for example, they might observe that they were ageing more or less rapidly than their onboard clocks would suggest. Everything appears normal to the astronauts on board the craft, but so far as Earthbound observers are concerned the crew are ageing more slowly than their terrestrial counterparts. If one member of a pair of twins were to make a long journey at close to the speed of light he would return to find his stay-at-home twin much older than himself. For example if, on their twentieth birthday, Jane were to set off on a voyage to a star 21 light-years away in a spacecraft which travels at 99% of the velocity of light, while her twin brother John remained behind then, neglecting time spent in acceleration and deceleration, she would return to Earth after just over 42 years of Earth-time to find John now aged 62. For Jane only about six years will have elapsed.

The result may appear ridiculous, but the time dilation effect has been confirmed in a variety of experiments, and there is no doubt that what has been described here is what would happen in a real situation. For example, cosmic rays (charged atomic particles reaching the Earth from space) strike the atmosphere and produce very short-lived particles called muons, which decay in an average period of about two millionths of a second measured in a frame of reference in which they are at rest. These particles are produced at an altitude of at least ten kilometres; even though they travel at nearly the speed of light, if there were no time dilation effect they would cover a distance of less than one kilometre before disintegrating. But, because of their very high speeds, the time dilation factor is sufficiently large for their lives to be extended to such an extent that they can be detected at ground level.

This explanation was put forward in 1941 by B. Rossi and D. B. Hall, and since that time many laboratory experiments on short-lived atomic particles have confirmed the predictions of Special Relativity. Indeed, in 1971, J. C. Hafele and R. E. Keating made a direct test by carrying atomic clocks around the world in jet airliners and comparing the times recorded with those which had elapsed on a "stationary" clock at the US Naval Laboratory; the results agreed with the theory.

Time dilation opens up the possibility of time travel—but only into the future. Travel into the past is not permitted.

Theory of Relativity that light is deflected in a gravitational field; stars close to the Sun in the sky were seen to be deflected from their normal positions by an amount consistent with Einstein's theory. (Courtesy Royal Greenwich Observatory.)

Sirius A and B. The bright star Sirius (the brightest star seen in our skies) has a much fainter companion (B) shown in this series of photographs. The two stars form a binary system. Although both stars may be seen when their angular separation is at its greatest, when the stars are close together the brilliance of Sirius A drowns out the light of B. The presence of B was deduced in 1834 as a result of observations of the gravitational influence which B exerts on the motion of A, but it was not until 1862, that B was observed for the first time. Sirius B is a white dwarf. (Lick Observatory Photograph.)

Mass increase: In adapting Newton's second law to fit the Special Theory of Relativity, Einstein found that a further consequence of the theory was that the mass of a body is affected by its motion. The mass of a moving body as judged by a "stationary" observer is greater than its rest-mass (i.e., the mass which it would have if it were stationary in the observer's frame of reference). The closer the body approaches the speed of light, the greater its mass becomes until, if it were possible for it to travel at the speed of light, its mass would become infinite. This implies that no material body can be made to travel at the speed of light, for the amount of energy required to accelerate by even a small increment a massive body which is already travelling close to the speed of light is very high; and an *infinite* amount of energy would be required to accelerate even the dot on this "i" to the speed of light. Given the resources, the speed of light may be approached as closely as we wish, but it can never be attained. The speed of light is an absolute barrier to the velocities of material particles.

The equivalence of mass and energy: Another consequence of the theory, related to the variability of mass, is that mass and energy are interchangeable: mass may be converted into energy, and energy into mass. If a certain amount of mass (m) is converted into energy, the amount of energy released (E) is given by the formula, $E=mc^2$, where c denotes the velocity of light. Since the velocity of light is a large number, and when squared is an even larger number, it follows that a great deal of energy may be obtained from the destruction of a small amount of matter. This aspect of the theory was vitally important, for it led us to understand how the Sun and stars are shining (see Chapter 6), and how to harness nuclear energy. It also led, alas, to the development of the fearsome array of nuclear weapons which exist in the world today. In this latter respect, Einstein commented: "If only I had known, I should have become a watchmaker."

Faster than light?

It is central to Relativity Theory that nothing can travel faster than light. In fact, this statement is not *quite* correct. Certainly no material object can be made to travel *at* the speed of light, and it therefore seems self-evident that nothing can travel *faster* than light; after all, if you are driving along and accelerate from 50km per hour to 70km per hour, at some point you must have been travelling at 60km per hour. However, it has been noted that one can consider particles which have finite values of mass and energy provided that they always travel faster than light: as their velocities decrease towards the "light barrier" their masses increase towards infinite values. Such hypothetical particles have been labelled *tachyons*; just as ordinary matter particles cannot be accelerated up to the speed of light, so tachyons cannot be slowed down to the speed of light. Whether or not tachyons exist is as yet a matter of debate.

The crucial point about relativity is that no information can be communicated faster than light. If it were possible to send a faster-than-light signal then the most bizarre consequences would ensue. It would be possible to know the outcome of an event before it took place, and then—armed with that knowledge—to prevent the event taking place! Clearly there is a paradox here. Again, if a space traveller could make a faster-than-light journey, he could return before he set off and persuade himself not to go on the grounds that he had already been!

If information could be communicated faster than light, then the fundamental law of causality, that cause must always precede effect, would be violated. There would be no logical pattern to events in the Universe, and events would be random and unpredictable. We have a vested interest in its *not* being possible to communicate faster than light. Even if tachyons exist, however, there need not be a problem, provided that they cannot be used to convey information from one place to another.

Space-time

In everyday experience we are used to thinking of a world of three dimensions. Objects in space have length, breadth and height. In tacit agreement with Newton we tend to think of time as something separate and independent, which flows past at a uniform rate. However, it is clear from Special Relativity that we cannot think of time as being detached and immutable. There is no absolute standard for the measurement of time or of space; both kinds of measurement are affected by the relative motion of observers. If two observers each in uniform relative motion were to see the same two separated events they would not agree either on the separation in space between these two events, or on the period of time which elapsed between them.

In 1907 the Russian-born mathematician Hermann Minkowski (1864–1909) proposed that we should treat the three dimensions of space and the dimension of time as being intimately linked; all events in the Universe should be as occurring in a four-dimensional *space-time*. As Minkowski remarked: "Henceforth space by itself, and time by itself, are doomed to fade away into mere shadows, and only a kind of union of the two will preserve an independent reality." The sum of all events Minkowski called "the world", and the path of an individual particle in space-time he called a "world-line". Clearly we cannot draw in four dimensions, but we can readily draw a "map" in which we measure time in the vertical direction and distance in the horizontal direction. Such a map is known as a *space-time diagram* (fig. 16). In the figure the vertical line represents a stationary particle, which stays for all time in the same place, the straight line leaning at an angle to the verticle represents a particle moving with uniform velocity, and the curved line denotes a particle which accelerates from rest to a finite velocity.

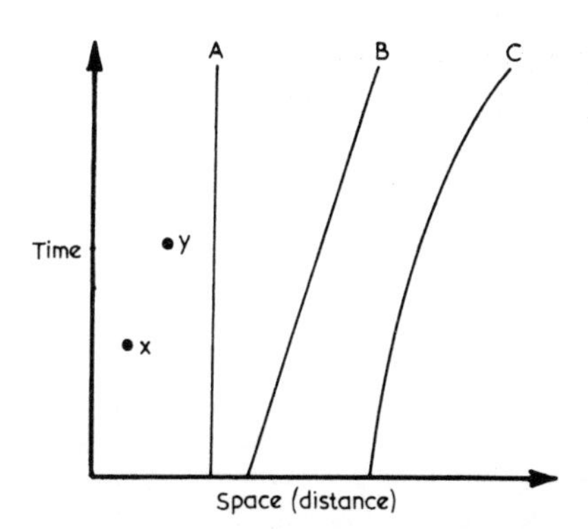

Fig. 16 *A space-time diagram.* Time is measured in the vertical direction and the three dimensions of space are represented by the horizontal direction; i.e., the horizontal direction denotes distance in space. x and y are two events, or world-points. Line A represents a stationary particle (its position in space does not change with time), line B denotes a particle moving with constant velocity (its position changes at a uniform rate) and line C denotes an accelerating particle (which starts from rest and moves with increasing velocity).

It is convenient to make the scales on the axes of these diagrams such that the speed of light is represented by a line which makes an angle of 45° to each axis. A ray of light travels a distance of some 300,000km in one second; therefore one second of time on the vertical axis has the same "length" as 300,000km on the horizontal scale. Since material particles must travel slower than light, their

(a)

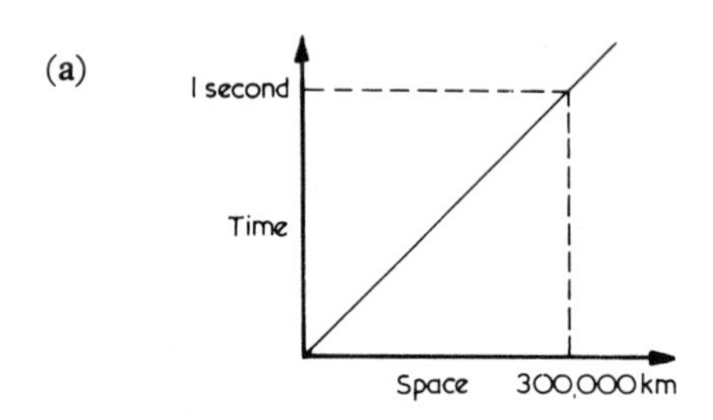

(b)

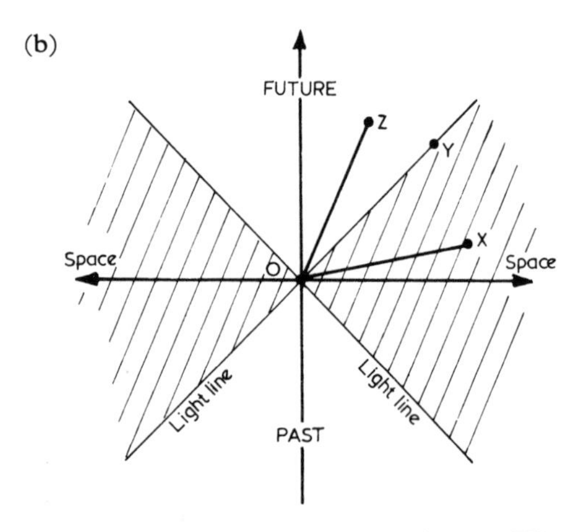

Fig. 17 *Accessible and inaccessible regions of space-time.* (*a*) The scale of the space-time diagram is often chosen such that the speed of light, 300,000km/sec, is represented by a line inclined to the vertical (time) axis by 45°. (*b*) The space-time diagram centred on observer O and extended into the past and future. The lines denoting the speed of light (light-lines) drawn through O divide all of space-time into three regions. The shaded region to left and right is inaccessible to O since to reach a point anywhere in it he would have to travel faster than light. The upper cone lying between the light-lines represents the future of O for, without exceeding the speed of light, he could be present at any point in that region or could influence any point within that region by means of a signal. The lower cone represents his past; again, he could have been present at any point in that region, or could have been influenced by a signal emanating from such a point. Line OX is called *space-like* and represents a forbidden, faster-than-light trip; OY denotes a speed-of-light trip, or *null* trajectory; and OZ denotes a *space-like*, slower-than-light trip.

world-lines must be inclined to the vertical by angles of less than 45°: such lines are called *time-like.*

The speed of light limit on velocity allows us to divide up the whole of space-time for a given observer into three distinct regions (fig. 17). The upper cone in the figure represents future events at which the observer can be present by travelling slower than the speed of light; this is his *future.* The lower cone represents past events at which he could have been present—again without exceeding light velocity. The rest of space-time is inaccessible to him. He cannot influence events in that region without making a faster-than-light trip, which is forbidden. Forbidden trajectories, lines inclined to the vertical by more than 45°, are called *space-like.* If relativity is correct, no information can be passed along space-like paths.

As we shall see in later chapters, space-time diagrams provide a very useful means of examining the behaviour of particles and material bodies.

One very useful property of space-time is that in its four dimensions we can define the (four-dimensional) *interval* between events in a way which will be agreed upon by all inertial observers. Although individual observers will disagree about the distance in space, and the period of time between two events, if they combine their observations of space and time in the appropriate way they will end up with exactly the same value of space-time interval between events. The space-time interval between events has a definite, absolute value. Different observers will measure different projections of this interval on their time axes and their space axes, but they will all agree on the value of the interval. Different observers in uniform relative motion will see different aspects of the interval, just as an oblong rectangular box looks like a rectangle when seen from one side and a square when seen from another.

Einstein was quick to realize the benefits of the space-time description as the framework for the treatment of Special Relativity, and the laws of nature have since been expressed in four-dimensional terms. Our Universe, then, seems to be four-dimensional. We cannot think of space and time as being separate entities; instead they are intimately linked together. If we alter our perception of space—for example, by travelling in a high-speed spacecraft—we alter our perception of time, too.

Special Relativity brought about a complete revolution in the way we think about space-time, and the Universe, but it was not the only revolution to take place in physics at the beginning of the twentieth century. At much the same time another equally significant upheaval was taking place in our understanding of the nature of radiation and matter.

Quantum theory

A "black body", in physicists' terms, is an ideal radiator of energy. It absorbs all radiation which falls upon it, and emits all wavelengths of radiation. The problem which arose at the beginning of the twentieth century was that the classical theory (based on Maxwell's equations) of the emission of radiation from hot bodies did not agree with observation. The theory, as derived around 1900 by Baron Rayleigh (1842–1919) and James Jeans (1877–1946) agreed well with observations at longer wavelengths, but at short wavelengths it predicted that all bodies having temperatures above absolute zero (—273° Celsius = 0K) should be radiating infinite quantities of energy. Clearly this could not be so—or we would not be here! Similarly, the German physicist Wilhelm Wien (1864–1928) had derived a theory which worked well for shorter wavelengths but not for longer ones.

In 1901 the German physicist Max Planck (1858–1947) resolved the problem by proposing that radiation is emitted in small packets, or *quanta*, of energy, the energy of each quantum being related to the corresponding wavelength of emitted radiation by a constant, now known as Planck's constant. When it was appreciated that radiation was released in a series of tiny, discrete "bursts", it was possible to construct a theory of radiation which agreed with observation and eliminated the drastic possibility (known as the "ultraviolet catastrophe") that all bodies would be radiating infinite amounts of lethal short-wave radiation.

In 1911, Ernest Rutherford (1871–1937) of the Cavendish Laboratory, Cambridge, proposed the theory that the atom—hitherto considered as the smallest indivisible entity of matter—consisted of a massive, central nucleus carrying positive electrical charge, surrounded by orbiting negatively charged "lightweight" particles known as electrons. Two years later the Danish physicist Niels Bohr (1885–1962) developed a theory of how atoms radiate energy (in the form of light and other electromagnetic radiations) based on quantum principles. He argued, in the case of the hydrogen atom, that the electron could move only in certain particular orbits, rather like the planets moving around the Sun. When the hydrogen atom receives energy the electron will jump up from a lower orbit (of lower energy) to a higher orbit (of higher energy). Conversely, after a brief interval, that electron drops back down again to a lower energy level: the energy released in dropping down to the lower level is emitted as a quantum of light of a characteristic wavelength, the wavelength being related to the change in energy of the atom by means of Planck's constant.

This discovery accounted for the fact that, when a rarefied gas of a particular chemical element (such as hydrogen) is heated up so that it emits light, that light is emitted at particular wavelengths only, not in a continuous spectrum of all colours and wavelengths. Likewise, if a continuous spectrum of light from a dense hot body (such as the interior of the Sun) is passed through a more rarefied gas, that gas will absorb light at those same particular wavelengths, so that when the light is examined with a spectroscope the resultant spectrum consists of a continuous band of colour superimposed upon which is a pattern of dark *absorption* lines.

The quantum view of matter and radiation was greatly supported by experiments which showed that electrons could be ejected from certain kinds of solids by shining light upon them. The curious thing was that the energies of the ejected electrons depended on the wavelength of light being used rather than the intensity of the beam. Einstein explained this, the photo-electric effect, in terms of quantum theory by pointing out that the energy acquired by an electron depended on the wavelength of the quantum of light which it absorbed. The shorter the wavelength, the greater the energy. Consequently, light of wavelength shorter than a certain value was necessary to dislodge an electron from its atom; a more intense beam of *longer* wavelength radiation would not achieve this effect. It was for this discovery that Einstein was awarded the 1921 Nobel Prize for Physics. Quanta, or "particles", of light are known as *photons*.

These and other discoveries showed that in some respects light behaved like particles, in others like a wave motion. A remarkable experiment carried out in 1927 by C. Davisson (1881–1958) and L. H. Germer (1896–1971) showed that a stream of electrons striking a crystal behaved in the way in which waves would be expected to behave. Particles, it seemed, sometimes behaved like waves and waves sometimes behaved like particles. Matter and radiation exhibited a wave-particle duality such that sometimes it was better to regard them as waves and at other times to treat them as particles.

When two beams of light meet up, they interfere with each other to produce a pattern of light and dark zones on a screen. The shape of the interference pattern can be calculated from the *wave* properties of light. We can also look at this from the point of view of photons (*particles* of light): the wave description tells us there is a high probability of finding photons in some locations (corresponding to the bright zones), and a low probability of finding them in others (the dark zones).

What quantum mechanics demonstrates is that the concept of *probability* enters nature at a very fundamental level. On the microscopic level (e.g., dealing with individual photons or particles) we cannot predict the result of a particular experiment (where that photon might land); all we can do is calculate the probabilities of different outcomes (the photon is more likely to land in one place than another). Only with large numbers of particles can we make reasonably precise predictions of how the experiment will go. This is a very profound viewpoint, for it implies that there is a limit to our ability to predict the future course of events. There is an inescapable element of chance in the Universe.

This aspect of quantum theory was made clear in 1927 by the German physicist Werner Heisenberg (1901–1976) in his *uncertainty principle*. This demonstrated that it was impossible to measure precisely two complementary properties of a particle, such as its velocity and position. By the very act of trying to observe one quantity (say, by trying to pin down position precisely) we affect the other quantity (velocity). Heisenberg illustrated this principle by reference to a hypothetical microscope. If we were to try to measure the position of an electron whose momentum was already known then, in order to see the electron and determine its position at a precise instant, we would have to illuminate it with photons. The energy transferred to the electron by a photon striking it would change its momentum by an unknown amount. We end up by knowing its position, but being uncertain about its momentum! This is a fundamental principle of crucial importance. It implies that, the more precisely we know one quantity, the less precisely we can know another related quantity. The amount of uncertainty in the knowledge of these quantities is related by Planck's constant, itself one of the fundamental constants of nature.

Great advances have ensued in subsequent years in the understanding of the nature of particles and in the application of quantum theory in a wide variety of branches of physics. The basic building blocks of matter, atoms, have been shown to be divisible into a bewildering array of subatomic and nuclear particles. The simple picture of an atom consisting of a heavy nucleus made up of positively charged protons and neutral neutrons, surrounded by a cloud of negatively charged electrons, has given way to a picture where material particles can be subdivided into a host of more fundamental particles characterized by various properties such as mass, charge and spin (electrons, for example, can be envisaged as "spinning" on an imaginary axis). It was shown in 1928 by the Cambridge physicist P. A. M. Dirac that each fundamental kind of particle has its antiparticle, a particle with mirror-image properties: for example, the antiparticle of the familiar electron is the positron, a particle which has the same mass as the electron but opposite charge and spin. If a particle meets up with its antiparticle, the two annihilate each other, releasing a burst of high-energy gamma radiation. Conversely, and in keeping with Einstein's equivalence of mass and energy, it has been shown that high-energy gamma rays can also create pairs of particles and antiparticles—electrons and positrons.

Quantum theory and Special Relativity have successfully been married into quantum electrodynamics, a theory of the electromagnetic force which treats the interactions governed by that force in terms of the interacting particles passing photons between each other. The forces which act in the nuclei of atoms, the so-called strong and weak nuclear forces, can also be described in quantum terms. Physicists feel that it should also be possible to describe gravitation in quantum terms, but so far they have met with little success.

Special Relativity and quantum theory turned our ideas of space, time, matter and radiation upside down in two revolutions which occurred in the early part of this century. Einstein was to be responsible for another revolution which took place in our conception of the nature of gravitation with the publication, in 1915, of his General Theory of Relativity.

5
Gravity Examined Anew

Newtonian gravitation was regarded as a force which acted instantaneously across space between massive bodies. The force of gravity acting between two bodies depended on their masses and on the distance between them. This posed no particular problems while absolute space and time were taken for granted, for the separation between the two bodies and the time could be defined unequivocally. Special Relativity swept away these two fundamental pillars of Newtonian theory: space and time were not absolute, and different observers in uniform relative motion would disagree on the question of the time and the separation between the bodies. Furthermore, since the speed of light represents an absolute barrier to the communication of information in the Universe, the idea of the instantaneous propagation of gravity was not acceptable.

Nevertheless, Newton's theory was a very good theory indeed, as its many successes had shown. If a new theory of gravity based on relativistic principles were to emerge, it would have to agree in its predictions with Newtonian theory in all but the most exceptional circumstances. By the beginning of the twentieth century there was only one example of a situation in which Newton's theory had failed—even then, only marginally—to account for an observed effect. The orbit of the planet Mercury had been shown by U. Leverrier in 1859 to behave in a way which could not fully be explained by Newton's law of gravitation. Due to the known perturbing effects of the other planets in the Solar System, the orbit of the planet should slowly move around the Sun in such a way that the point of closest approach (perihelion) advances in position by an angle of 5557 seconds of angular measurement per century. In fact, as Leverrier pointed out, the perihelion of Mercury advances by an amount greater than the expected value by 43 seconds per century.

Not a very large discrepancy, perhaps, but such was the level to which gravitational theory had advanced by then that the perturbing effects of the planets could be calculated to much better precision than the size of the error. Leverrier suggested the discrepancy might be due to the presence of a hitherto unknown planet closer to the Sun than Mercury. Such a hypothetical planet, tentatively named Vulcan, has never in fact been discovered. The puzzle was not to be resolved until 1915, when Einstein published his General Theory of Relativity.

Einstein's principle of equivalence

The first key foundation of that theory was laid in 1907 when Einstein formulated his principle of equivalence. This is a development of the earlier principle established by Galileo that all bodies, whatever their masses, accelerate at the same rate when falling under the influence of a gravitational field, the implication being that inertial mass and gravitational mass are exactly equal. The equivalence of gravitational and inertial mass, as we saw in Chapter 3, has been tested to an accuracy of about one part in a million million. Why these two types of mass should be equal has been regarded as a mystery; the fact that they *are* equal, and that all bodies fall at the same rate in a gravitational field, is sometimes known as the *weak* equivalence principle.

Einstein pointed out that an observer inside a closed box cannot distinguish between the effects of gravitation and those produced by acceleration. An observer in a box on the Earth's surface (fig. 18) will feel his normal weight, and will note that all falling bodies accelerate towards the floor at exactly the same rate. If this observer were placed in a similar box in the depths of space, and if the box were fitted with a rocket motor capable of accelerating it smoothly at a rate exactly equal to the acceleration due to gravity* here on Earth, he would again find that all free objects accelerated towards the floor at exactly the same rate. He would also feel his normal

* Objects falling freely near the surface of the Earth accelerate at a rate of about 9.8 metres per second, each second, so that after 1 second a body will have reached a speed of 9.8 metres per second, after 2 seconds 19.6 metres per second, and so on. This figure is known as the acceleration due to gravity, and is denoted by the symbol g.

weight. There are no observations or experiments which he could carry out within the confines of his closed box which could tell him whether the effects were of gravity or of acceleration. Within a small closed box the effects of gravity and acceleration are indistinguishable.

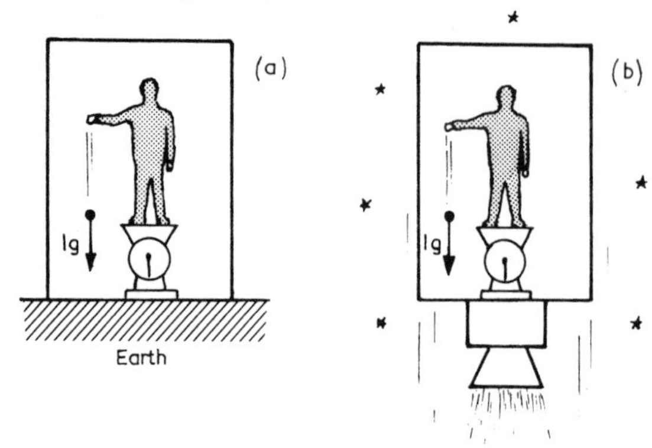

Fig. 18 *Equivalence of gravitation and acceleration.* An observer in a closed box cannot tell whether (*a*) he is standing on the surface of the Earth or (*b*) is out in space, accelerating at a uniform rate equal to the acceleration due to gravity at the Earth's surface.

Gravity in this context behaves like an *inertial* force, a "fictitious" force which arises due to the acceleration of the frame of reference from which observations are being made. The most familiar example of an inertial force is "centrifugal force". If an observer inside a closed van is being driven along a smooth straight road at a steady speed he is unaware of any forces (other than his weight) acting upon him. However, as the van turns a corner he finds himself thrown against the side of the van by what, to him, seems to be a perfectly real force. To an outsider, however, things are seen in a different light. What has happened is that, in accordance with Newton's first law, the observer has continued to move in a straight line at uniform speed. The van—his frame of reference—is accelerating, by turning the corner, and the act of turning the corner has caused the side of the van to collide with the observer. No force has accelerated the observer and thrown him against the side of the van; that impression was given by the accelerating frame from which his observations were made.

If the effects of gravity and those of acceleration are indistinguishable, is there a sense in which we can regard gravity as an "apparent force"?

Imagine a closed box again—an elevator this time (fig. 19). If the supporting cable were to snap, the lift and its contents would fall freely under the influence of gravity, everything accelerating at precisely the same rate. The observer would no longer feel any sensation of weight, and objects released inside the elevator would float freely without any tendency to accelerate towards the floor. Everything inside the elevator would be weightless. From an outsider's point of view everything inside the elevator would be accelerating at precisely the same rate as the elevator itself, and there would be no acceleration of any of the contents relative to the floor of the lift. The observer inside the box would be quite unable, by any experiment undertaken inside the box, to tell whether he was falling down the elevator shaft or floating freely in space.

Fig. 19 *Free fall.* The observer, contained in a stationary elevator, weighs 70kg (*a*). If the cable snaps, the elevator and its contents fall freely under the attraction of the Earth's gravitational field (*b*); since the elevator and everything inside it are accelerating at the same rate, the observer no longer feels himself pressed against the floor by gravity, and so experiences the sensation of weightlessness. There is no way that the observer could tell by experiments carried out inside the box (elevator) whether he was falling down the shaft or floating freely in space far from the gravitational influence of the Earth (*c*).

From these examples we see that the effects of gravity can be created or eliminated altogether by choosing a suitable frame of reference.

Newton's laws of motion would hold good inside a freely falling box. For example, if a body were set in motion, it would continue in that state of motion (until it hit the side of the box) in accordance with the first law. It is easy to see that the other two laws would be obeyed as well. The freely falling box, then, constitutes a *local* inertial frame: within the confines of the box all the conditions required to define an inertial frame (Chapter 3) are met. Einstein's equivalence principle stated not only that within closed boxes the effects of gravitation and acceleration were indistinguishable, but that all the laws of nature have the same form within freely falling boxes as they would have in any inertial frame of reference. In this formulation, sometimes known as the *strong* equivalence principle, the implication is that all freely falling boxes are equivalent for the performance of all physical experiments.

It is important to note that the equivalence principle holds true only within volumes of space small enough for gravity to be regarded as constant. If the box is too large, tidal effects will be observed—the floor of a box falling towards the Earth would be nearer to the centre of the Earth than would the top of the box, and so a particle released near the top of the box would accelerate less rapidly than one released near the floor; as a result the two particles would drift slowly apart. Einstein extended the concept of an inertial frame to include any and all freely falling laboratories, and he removed from the concept its tacit association with absolute space (in Newton's view an inertial frame was one moving uniformly relative to absolute space) or with the frame of reference made up of the distant galaxies. He also restricted the concept to being a local one: because the effects of gravity are present everywhere in the Universe, and the strength of gravity differs from place to place depending on the distribution of matter, in large freely falling frames of reference differential effects—such as the tidal effect mentioned above—will become apparent and such frames cannot be regarded as being truly inertial (bodies initially at rest will drift apart in contravention of Newton's first law).

Consequences of the equivalence principle

If the effects of gravitation and acceleration are indistinguishable, then rays of light should be deflected in a gravitational field, and light moving up through a gravitational field should be red-shifted. Let us look at these phenomena in turn.

Consider again our ill fated observer in a freely falling elevator. By the principle of equivalence no effects of gravity should be apparent to the observer in the elevator, so that if he throws a ball towards the opposite wall it will travel in a straight line (fig. 20). Relative to observers standing outside, the elevator is accelerating downwards, and so the ball is seen to fall with it, describing a parabolic path just as any projectile would do when thrown just above the Earth's surface.

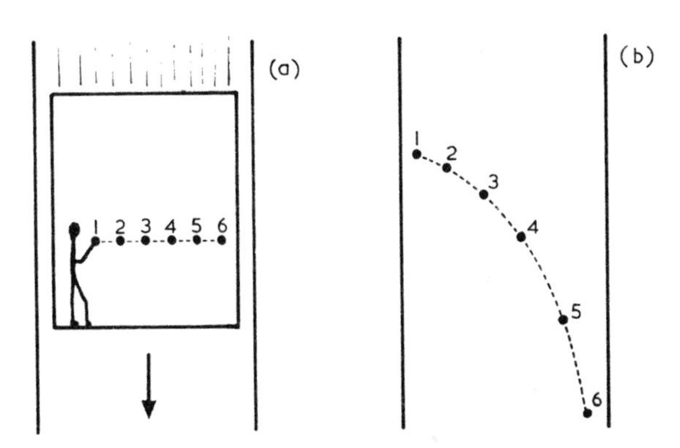

Fig. 20 *The relativity of free fall.* (*a*) If the observer inside the freely falling box throws a ball towards the opposite wall, it will travel in a straight line, in his frame of reference. (*b*) To an outside, "stationary" observer, the elevator and its contents are accelerating downwards under the influence of gravity; accordingly, from his point of view the ball will follow the parabolic path associated with a projectile moving in the Earth's gravitational field (cf fig. 9, page 40).

If, instead, the observer shone a beam of light across the cabin, it too would follow a straight-line path within the elevator. From the outsider's viewpoint that ray of light would follow a curved path, for in the (very short) time taken for it to cross the elevator, the elevator will have fallen further down the shaft. The curvature would be very slight since the velocity of light is very great, but there is no doubt that the path of the light ray should be bent.

The conclusion to be drawn from the equivalence principle is that rays of light should be deflected when passing close to massive bodies. In fact, the equivalence of mass and energy which emerges from the Special Theory suggests that light should be affected by gravity just as matter is, and so this result should not be too surprising. As we shall see, the bending of light by gravity has been confirmed by experiment.

In 1842 the Austrian physicist Christian Doppler (1803–1853) showed that the motion of a source of sound waves would affect the frequency of the waves (the pitch) received by a "fixed" observer.

We are all familiar with how, when a source of sound such as a siren mounted on a moving vehicle is approaching us, the pitch of the note which we hear is significantly higher than when that source is receding.

According to the *Doppler effect*, if a source of light is receding from an observer the number of wavecrests per second which he receives (i.e., the frequency) will be less than the number which he would receive if that source were stationary relative to him. In effect, the light waves are "stretched out" because of the recession of the source (fig. 21), and so the perceived wavelength is longer. Since red light corresponds to the long-wave end of the visible spectrum, this effect is called the *red-shift*. Similarly, if a source is approaching, the light received by the observer is blue-shifted.

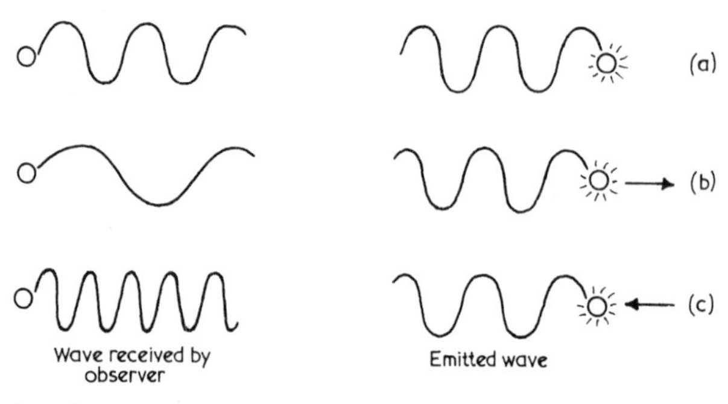

Fig. 21 *The Doppler effect.* (*a*) A ray of light received by an observer from a source stationary with respect to him will arrive with the same wavelength as the emitted wave. (*b*) If the source is receding from the observer, each successive wave crest is emitted from a slightly greater distance than the preceding one, and so takes marginally longer to reach the observer; accordingly the number of wavecrests per second reaching the observer is fewer than the number emitted at the source, and the perceived wavelength is longer than the emitted wavelength. (*c*) Conversely, if the source is approaching the observer, the light waves are "squeezed up" and the perceived wavelength is shorter than the emitted wavelength.

As the French physicist A. H. Fizeau (1819–1896) showed in 1848, this effect will result in a change in wavelength of the emission or absorption lines present in the spectrum of a star which is approaching or receding from an observer, the change in wavelength ($\Delta\lambda$) compared to the "true" wavelength (λ) being dependent on the velocity (v) of the source relative to the observer. Red-shift is defined as $\Delta\lambda/\lambda$ and, for velocities which are not a large fraction of the speed of light (c), is equal to v/c. Hence observations of the red-shift (or

Supernova in the galaxy NGC 7331. The brilliance attained by an individual supernova can be appreciated by comparing the pictures taken before and during the maximum brilliancy in 1959; the position of the supernova is indicated by the arrow. (Lick Observatory Photograph.)

The Crab Nebula. This tangled expanding cloud of gas, a source of all kinds of electromagnetic radiation from X-rays to radio waves, is the remnant of a supernova (exploding star) which occurred in the year 1054. At the heart of this object lies a pulsar, powered by a spinning neutron star, and 'flashing' about thirty times per second. (Photograph from the Palomar Observatory, California Institute of Technology.)

X-ray picture of the Crab Nebula taken by the Einstein Observatory (HEAO 2) in 1979. The bright object at the centre is the pulsar, which is transferring energy to the surrounding medium. (Courtesy NASA.)

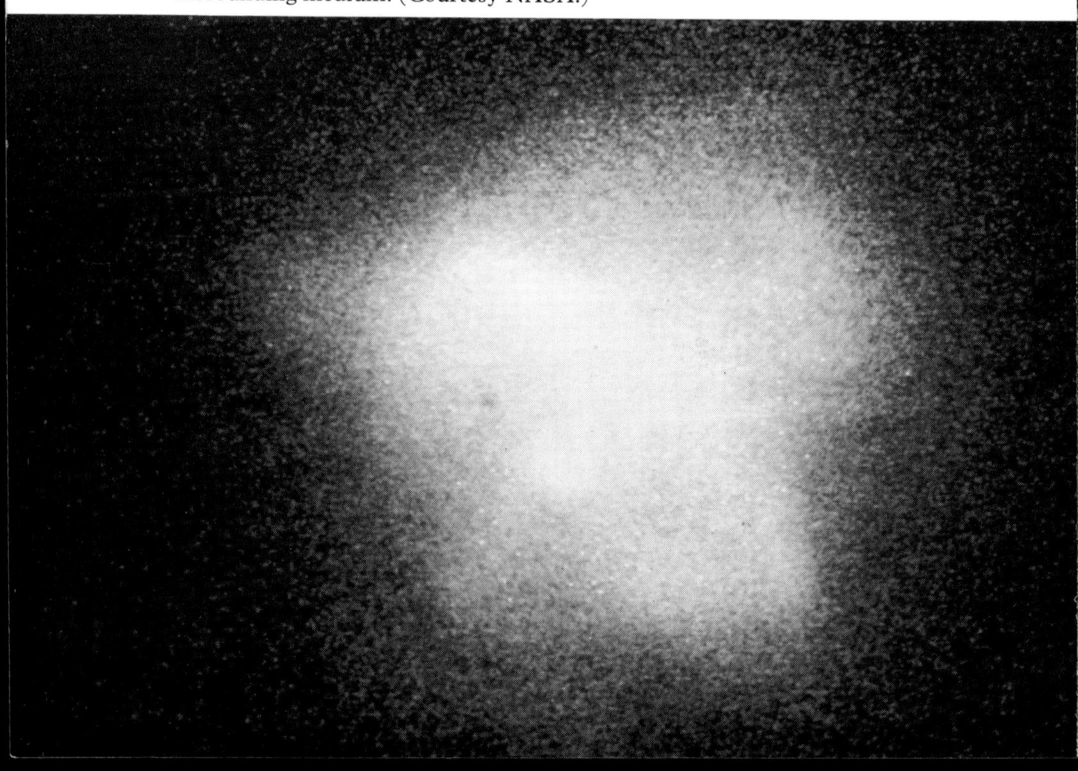

blue-shift) can yield the velocity of recession (or approach) of the source.

The equivalence principle shows that a red-shift should also occur in light moving up through a gravitational field. Consider our observer in the freely falling elevator. If he shines a beam of light upwards from the floor then, by the equivalence principle, there should be nothing about this beam of light which would indicate to him that his lift was falling down a shaft rather than floating freely in space. However, to a fixed observer looking down the elevator shaft the source of light will be receding, and so he will receive red-shifted light. The conclusion to be drawn is that light moving upwards through a gravitational field will be red-shifted (light falling down will be blue-shifted).

Curved space-time

The other key element in the formulation of general relativity was the concept of curved space-time. The space-time of Minkowski, which was the space-time of Special Relativity, was *flat*: the shortest distance between two points was a straight line, and the angles of a triangle added up to 180°. As we have just seen, light is the fastest-moving entity in the Universe, yet even light is constrained to follow curved paths in the presence of matter. Since the gravitational effects of matter are present, however weakly, everywhere in the Universe, no particle, whether a photon of light or a lump of rock, can follow a truly straight-line path in space. Particles of matter are subject to accelerations as they pass massive bodies, and of course the world-lines of accelerating particles are curved.

We have seen that it is possible to produce or eliminate the effects of gravity by choosing a suitable frame of reference, and this makes us suspicious of the nature of gravity as a "force". Suppose that the effect of matter were to distort the geometry of space so that, in the vicinity of matter, space-time was no longer flat, but curved. If that were so then the shortest distance between two points would not be a straight line, and the paths of rays and particles near matter would be curved. In such a conception, gravitation would not be a force, acting directly between individual masses. Instead, what we take to be a gravitational force would arise simply from the geometry of space-time.

Although this may seem a peculiar way of looking at gravity, it is easy to draw an analogy which may be helpful. Imagine two flat creatures living on the surface of a sphere (fig. 22). We have to imagine that these creatures are truly flat—two-dimensional creatures who can move forwards, backwards, left and right, but to whom the vertical direction has no meaning whatsoever (we shall meet them again from time to time, as they are rather useful). Suppose that these two creatures, A and B, set off at the same rate along parallel paths at right angles to the "equator" of their sphere, starting out from different points on the equator. We have to assume that these creatures, being flat themselves, assume their "universe" to be flat also, and to obey the laws of Euclidean geometry—certainly they could not possibly visualize the appearance of a sphere.

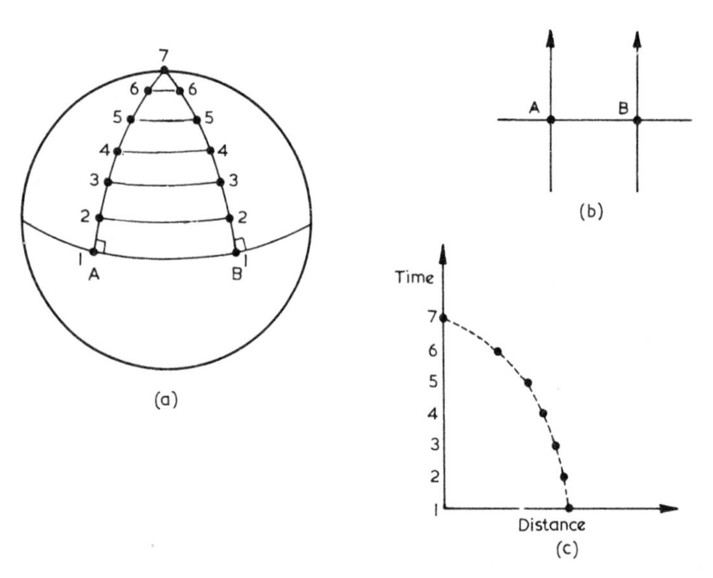

Fig. 22 *Geometric "forces" on the surface of a sphere.* (*a*) Two flat creatures, A and B, set off on parallel paths, perpendicular to the equator of the sphere on which they live—(*b*) by definition, their paths *are* parallel, for they make equal angles with the equator. A and B eventually meet at the pole, their separations at 7 different times being shown. From A's point of view, B accelerates towards him (*c*). A and B might conclude that a force of attraction had brought them together (for the vertical direction has no meaning for them), but we, as outsiders, can see that their coming together was merely a consequence of the geometry of the "universe" in which they are located.

As they progress on their journeys, A and B find that, slowly at first, and then more rapidly, they are being drawn inexorably together until they collide with each other at the "north pole" of their universe. Since, in order to avoid hitting each other, they each would have to exert a force directed away from the other, they might conclude, quite reasonably, that they were being drawn together by a force, a force which they might choose to call "gravity". It would be misleading to take the analogy too far, but it does demonstrate how a geometrical effect can give rise to a phenomenon which appears to be a force.

The geometry of positively curved space (space curved in roughly the same way as the surface of a sphere) was first investigated in 1854 by the German mathematician G. F. B. Riemann (1826–1866). In such a space there could be no truly "parallel" lines; all pairs of lines would eventually meet (just as, for example, lines of longitude on the Earth's surface meet at the poles). Just as on the surface of a sphere, in positively curved space the angles making up a triangle add up to more than 180°. In curved space the "shortest distance" between two points is a curved path known as a *geodesic.*

Navigators on the Earth are familiar with the fact that the shortest distance between two points on the surface of a sphere is not a straight line. The shortest route from A to B is to follow a *great circle*—that is, a circle whose centre is also the centre of the Earth; by contrast, a *small circle,* such as a "parallel" of latitude, lies in a plane which does not pass through the centre of the Earth. The shortest distance between two widely separated points which lie at the same latitude lies along the great-circle route between these points rather than along the circle of latitude on which they are located. By way of example, to travel along a circle of latitude between two points at latitude 60° N lying on opposite sides of the Earth involves a journey of about 10,000km. Taking the great-circle route directly over the pole involves travelling only about 6,700km.

As the next crucial step in formulating the General Theory of Relativity, then, Einstein proposed that space-time (not space alone) would be curved in the presence of massive bodies, and that rays and material particles would travel through space-time in the most efficient way possible—along geodesic paths. Gravitation results from the geometry of space-time in the vicinity of massive bodies. The paths of photons and material particles alike depend on the curvature of the space-time in which they are moving. For example, a planet travels in its orbit around the Sun, not because it is constrained so to do by a force of attraction acting directly between it and the Sun, but because it is responding to the distortion of space-time in the presence of the massive Sun.

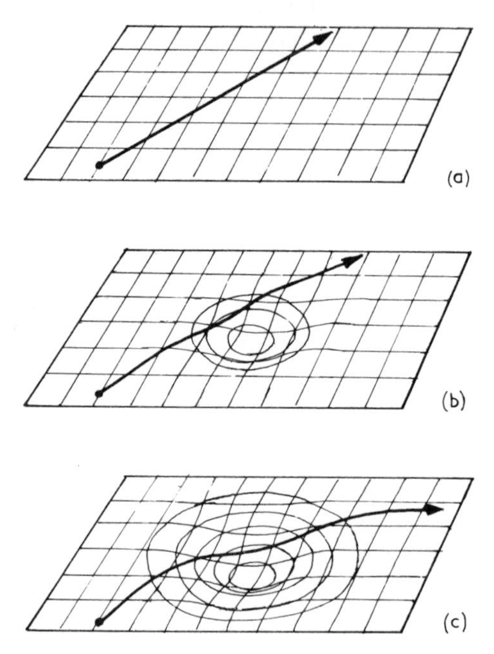

Fig. 23 *The rubber-sheet analogy*. Imagine space-time to be a rubber sheet. (*a*) On a flat surface a ball set in motion will move in a straight line; likewise, in the absence of matter, space-time is "flat" and particles and rays follow straight-line paths. (*b*) If a weight is placed on the sheet it causes an indentation which deflects the motion of a ball; likewise the effect of a lump of matter is to distort space-time so that rays and particles follow curved paths. (*c*) A greater weight causes a greater indentation; likewise a more massive body produces a greater distortion of space-time (i.e., it has a more powerful gravitational field).

The more massive and concentrated the body, the greater the curvature of space-time in its vicinity, and the greater the "force" of gravity experienced by bodies in its locality. With increasing distance from a massive body the curvature of space becomes much less (and the perceived gravitational "forces" correspondingly weaker), and in the absence of matter space-time would be flat.

The behaviour of space-time is often illustrated by the "rubber-sheet" analogy. In the absence of matter space-time can be represented by a flat rubber sheet (fig. 23), and a ball set rolling across this sheet would move along a straight line path. If a weight is placed on the sheet, it causes an indentation which will deflect the ball from its straight path; a heavier weight (representing a greater mass) causes a larger indentation, affecting the curvature of a larger area of the rubber sheet and producing a correspondingly greater deflection of the ball. A ball given the right velocity could remain in orbit in the indentation caused by the heavy weight, just as a planet remains in its orbit in the curved space-time surrounding the Sun. As with all analogies, this should not be taken too far, but it serves to illustrate how the different space-time curvatures associated with different masses would affect the world-lines of particles.

The General Theory of Relativity

The General Theory, published by Einstein in 1915, brought together the preceding strands. It incorporated the principle of equivalence and the concept that space-time is curved in the presence of massive bodies—indeed, it was the basic fact that gravitational and inertial forces are indistinguishable within closed boxes that pointed to the possibility of treating gravitation as an effect of the geometry of space-time rather than a force acting directly between massive bodies. The curvature of space-time arising from the gravitational field of matter is calculated by means of the *field equations*, and the paths of rays of light or material particles in curved space-time are obtained by solving the *geodesic equations.* I think it was Professor A. Wheeler who first described General Relativity most succinctly by remarking: "Matter tells space how to curve, and space tells matter how to move."

General Relativity radically changed our ideas on space, time and gravitation. Instead of being a force which acted at a distance, as in Newton's formulation, gravity became absorbed into the geometry of space-time. Bodies are not acted upon directly by forces; instead they respond to the curvature of space-time in their vicinity. Any change in the gravitational field of a body is not communicated instantaneously, but travels out through the field at the speed of light so that, if the Sun were to be annihilated at the instant you read these words, just over 8 minutes would elapse before the Earth "realized" that the Sun's gravitational "attraction" had vanished.

Where gravitational fields are weak, General Relativity and Newtonian gravitation give the same answers and make the same predictions about the motions of particles. Since gravity is a very weak force, under normal circumstances the differences between the predictions of the two theories are too small to be detected. However, in strong gravitational fields differences should become apparent, and it should be possible by observation or experiment to distinguish between the theories and decide which gives the better description of nature.

Tests of General Relativity

There are three "classical" tests of the theory. The first of these concerns the advance of the perihelion of Mercury, an effect mentioned at the beginning of this chapter. Einstein found that according to his theory the orbit of a planet moving around the Sun in the absence of perturbing effects due to the other planets should be a slowly precessing ellipse. The effect would be greatest for Mercury, and in that case would amount to 43 seconds of arc per century, a figure exactly equal to the discrepancy between the observed motion of that planet and the predicted motion based on Newton's theory. When the perturbing effects of the other planets were included in the calculation, General Relativity accounted completely for the observed behaviour of Mercury.

While it was a great success for the theory to account for a known discrepancy in this way, a more stringent test of a theory's value is in the prediction of effects which previously had not been suspected. General Relativity succeeded here in two significant aspects: the gravitational bending of light, and the gravitational red-shift, both of which, as we have already seen, arise as a consequence of the equivalence principle.

According to General Relativity, a ray of light passing the edge of the Sun should be deflected by an angle of about 1.75 seconds of arc (fig. 24): as stars could not (at that time) be observed close to the Sun in daylight, such a test had to be conducted during a total solar eclipse. The bending of starlight was examined in 1919 during the total eclipse of 29 May by a British expedition headed by Arthur Eddington (1882–1944). Analysis of photographic plates taken by the expedition showed clearly that apparent stellar positions were indeed displaced by the gravitational field of the Sun, and the value of the displacement obtained was 1.98 ± 0.18 seconds, a value comfortably close to Einstein's prediction. Coming so soon after the publication of the theory, these measurements provided compelling evidence in its favour. A subsequent remeasurement of the original photographic plates carried out in 1979 by G. M. Harvey of the Royal Greenwich Observatory yielded a value of 1.87 ± 0.13, in even closer agreement with the Einstein prediction.

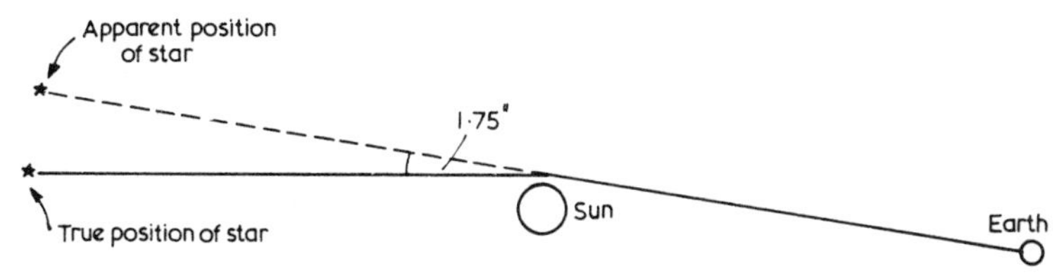

Fig. 24 *The bending of light by the Sun.* According to General Relativity a ray of light passing close to the edge of the Sun should be deflected through an angle of 1.75 seconds of arc. Seen from the Earth, the star will be deflected from its normal position by that angle.

Observations of this kind have been carried out at many total eclipses since and, although the inherent difficulties have led to varying results, nevertheless the bending of light by the Sun has been amply confirmed. The development of very long baseline interferometry (VLBI) techniques in radioastronomy, whereby the signals received by widely separated radiotelescopes can be combined to give very precise positional measurements, have allowed the bending of radio waves by the Sun to be measured with much greater precision. Every year the Sun passes close to a number of cosmic radio sources such as the quasars (see Chapter 10) 3C273 and 3C279, and observations of these have confirmed Einstein's prediction to within about 1%.

The third test concerns the gravitational red-shift which, as we saw, arises as a consequence of the equivalence principle. We can look at this effect in another way. Light escaping from the gravitational field of a massive body has to "work hard" to climb out of the gravitational well from which it sets out. In so doing, it loses energy and, as we saw in the previous chapter, the lower the energy of a photon, the longer its wavelength: consequently light emitted from a massive body should be red-shifted. An increase in wavelength corresponds to a decrease in frequency; the number of wavecrests passing a fixed observer in one second decreases as the wavelength increases. If we think of each wavecrest as corresponding to a "tick" of an atomic clock, then the frequency of the ticks from such a clock will be reduced in a strong gravitational field. The gravitational red-shift can be regarded as a gravitational time dilation effect. All clocks, natural, biological and manmade, will run more slowly in a strong gravitational field than in a weak one. The effect is very tiny indeed in the case of a body such as the Earth: the rate of a clock far from the Earth should differ from the rate of a clock at the Earth's surface by only seven parts in ten billion; i.e., the Earth-based clock should run slow by about 20 seconds in a thousand years. However, as we shall see in later chapters, the effects in strong gravitational fields can be dramatic indeed.

Although measurements taken in 1925 by the US astronomer Walter Adams (1876–1956) showed a red-shift in the light from the very dense companion star of Sirius, the first really convincing detection of gravitational red-shift/time dilation was achieved in 1960 by means of an ingenious experiment carried out by R. V. Pound and G. A. Rebka of Harvard University. They measured the frequency shifts which occurred in gamma-ray photons travelling up and down through a height of 23m (74ft) in a laboratory tower block, and found the frequency shift to agree with Einstein's predictions to within an accuracy of about 1%. Various different experiments have been undertaken in recent years to check this effect, one of the most interesting approaches being to fly high-precision atomic clocks around the world in opposite directions, so as to separate effects due to velocity (Special Relativity) from those due to the slight decrease in gravitational field experienced by the clocks while in flight some 10,000m above the ground (General Relativity). This was first accomplished in 1971 by J. C. Hafele and R. E. Keating, although the gravitational time dilation in that particular instance was confirmed only to within 10%.

A much more precise experiment was carried out in 1976 by R. F. C. Vessot and M. W. Levine of the Harvard-Smithsonian Center for Astrophysics. A hydrogen maser clock was fired in a rocket to an altitude of about 10,000km and the signals from this clock compared with those of identical clocks on the ground. Although the expected difference in the frequencies of the clocks was only about 4.5 parts in ten billion, the clocks were considered to be accurate over the period of the experiment to about one part in a thousand million million (i.e., 1 in 10^{15}). Although analysis of the results proved to be a lengthy and difficult task, by 1978 the experimenters were confident that their results agreed with the predictions of General Relativity to within two parts in 10,000.

In all tests to date General Relativity has shown itself to be superior to Newton's theory of gravitation, correctly predicting effects which do not arise in Newtonian theory. A number of alternative theories have been proposed since 1915, but none has proved itself to be superior to General Relativity so far, and some have already been eliminated in the face of experimental evidence. Later we shall examine some of these theories, other tests which have been devised to decide between them, and future tests to which General Relativity may be subjected. For the moment though, it is firmly installed as the best available theory of gravitation, and there it will remain until such time as it fails to withstand some new experimental test.

Although General Relativity was shown to be superior to Newton's theory early in the twentieth century, for a long time research into gravitational theory languished. In normal circumstances gravity is such a weak force that effects peculiar to General Relativity cannot be discerned, and physicists rely on Newtonian gravitational theory for their calculations; after all, Newton's theory is much easier to use. General Relativity comes into its own only when dealing with the Universe on the large scale, or with very powerful gravitational fields associated with massive concentrated bodies. During the nineteen-sixties and -seventies astrophysicists began to make observations in both of these areas which led to a sudden upsurge of interest in gravitational astronomy. On the one hand, astronomers discovered objects which were the highly compressed collapsed remnants of massive stars (neutron stars) and began to amass evidence favourable to the possible detection of the ultimate collapsed state of matter—black holes. On the other, observational evidence clearly favoured one particular type of cosmological theory, the Big Bang theory, whereby the Universe originated in a great explosive event a finite time ago; and observations were reaching a level of sophistication capable of revealing the rôle of gravity in the evolution of the Universe as a whole. In addition there were discovered classes of galaxies, and peculiar objects such as quasars, which showed evidence of containing compact yet immensely powerful energy sources. Many of these sources were considered to involve gravitational fields so powerful that only General Relativity would be adequate to deal with them.

Suddenly, or so it seemed, astrophysicists found that they had to draw more and more on Einstein's theory to deal with phenomena in the depths of the Universe. Conversely, astrophysics became the new testing ground for General Relativity. The results are proving to be exciting.

PART TWO

GRAVITY, BLACK HOLES AND THE UNIVERSE

6
Gravity and the Stars

The first successful attempt to demonstrate that the stars making up a binary revolve according to Kepler's laws and the predictions of Newtonian gravitation was made in 1827 by Felix Savary (1797–1841).

The two stars making up a binary move around their common centre of mass under their mutual gravitational attractions, but it is often convenient to regard one of the stars as fixed and the other as pursuing an elliptical orbit around it. From Newtonian theory we find that if we regard one star, of mas M_1, to be fixed, the companion star, of mass M_2, travels around M_1 as if it were a "planet" moving under the influence of a mass equal to M_1+M_2 located at the position of M_1.

If the orbital period in years (P) of the stars is known, and if the mean separation in astronomical units (a) of the stars can be determined, the combined mass of the two stars—expressed in units of the mass of the Sun—is readily calculated from the simple expression:

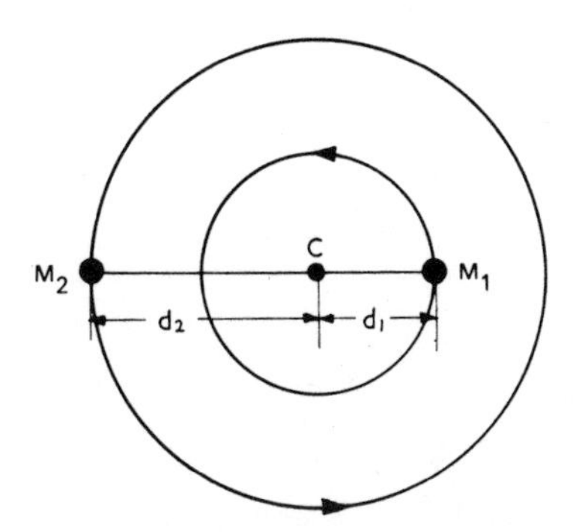

Fig. 25 *Centre of mass of a binary.* Both bodies, M_1 and M_2, revolve around their common centre of mass, C. C lies closer to the more massive of the two such that $d_1/d_2=M_2/M_1$.

$M_1 + M_2 = a^3/P^2$. This is simply Kepler's third law applied to two massive bodies rather than to the Sun and a planet.

In order to find the individual masses of the stars, it is necessary to find the relative distances d_1 and d_2 of the masses M_1 and M_2 from the centre of mass of the system (fig. 25). Imagine a weightlifter's bar with unequal weights at each end: if we wished to balance it we would have to hold it at a point closer to the heavier weight. In exactly the same way, we find that the ratio of the masses of the two stars in a binary is given by $M_1/M_2 = d_2/d_1$. Knowing the combined mass, and the ratio of the masses, the masses of the individual stars may be calculated. This is the only direct method by which the masses of stars can be obtained; it is difficult to operate in practice, and it is surprising how few precise measurements of stellar masses have been made.

In some binaries, known as astrometric, the presence of an invisible companion, too faint to be seen against the brilliance of the other, can be inferred from the "wobbling" motion of the brighter star as it moves through space under the influence of its fainter companion (fig. 26). According to Newton's first law, a single star should move through space at uniform speed in a straight line (ignoring the general gravitational field of the Galaxy); likewise, the centre of mass of a binary system will pursue a constant velocity through space. As may be seen from the figure, the two stars will weave to and fro about this line.

It was from such studies that the presence of an invisible companion to Sirius, the brightest star in the sky, was deduced by the German astronomer Friedrich Bessel (1784–1846). Although Bessel became aware of the wobbling motion of Sirius in 1834, it was not until 1862 that the US astronomer and telescope maker Alvin Clark

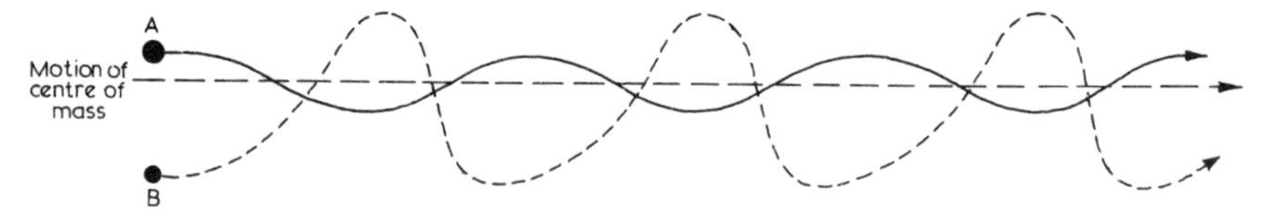

Fig. 26 *Astrometric binary.* Stars A and B make up a binary, A being more massive and far more luminous than B; B cannot be seen directly from the Earth. The centre of mass of the binary system follows a straight-line path through space while, due to their relative orbital motion, stars A and B move from side to side of this line. A will therefore be seen to follow a weaving path, which implies the presence of an invisible companion.

(1832–1897) succeeded in observing the faint companion, which turned out to be the first of a class of tiny, hot, dense stars known as white dwarfs.

Only a few nearby stars display sufficiently great apparent motion on the sky for such effects to be readily detected. Yet, despite the difficulty inherent in making these observations, there are a few cases where the evidence suggests that the gravitational influences of massive *planets* on their parent stars have been observed. The best example is Barnard's Star, a dull red star lying at a distance of just under 6 light-years which—according to a long series of observations carried out by Peter van de Kamp of Sproul Observatory—has two planets comparable in mass with Jupiter (it may have, of course, any number of smaller ones).

In most binary systems the stars are too close together to be seen as individual objects, and all that can be seen through a telescope is a single "star" made up of the combined light of the two components. Very often in such a situation the spectrum of their combined light reveals the presence of two stars for, as the stars revolve around each other (fig. 27), at any particular instant one of the stars is approaching and the other receding: in accordance with the Doppler effect, the spectral lines from the approaching star are blue-shifted while those from the receding star are red-shifted. Thus the spectral lines from each star will oscillate in wavelength in a periodic fashion, so betraying the presence of two stars.

Observations of this kind allow us to determine the period of the binary, and to obtain some information about the speeds of the stars in their orbits. However, our measurements give only the *radial* velocity (velocity towards or away from us), and the relation of this to the actual orbital velocities depends on the angle at which the orbit is tilted to our line of sight. It is usually possible to draw on other kinds of information to place limits on this angle of inclination, with the result that, from a knowledge of period and an approximate knowledge of orbital velocity, reasonable estimates can be made of the combined mass of the stars involved.

If the orbit should happen to lie edge-on or nearly so, then each star alternately will pass in front of the other, giving rise to an apparent variation in brightness of the visible "star". *Eclipsing binaries*, as these are called, can yield a great deal of information, as the angle of inclination is reasonably well known, and the duration and nature of the eclipses provides information on such factors as the sizes of the stars involved.

Gravity, then, provides the means of "weighing" the stars. And it plays another important rôle in binary systems, since it allows mass to be transferred from one component to the other.

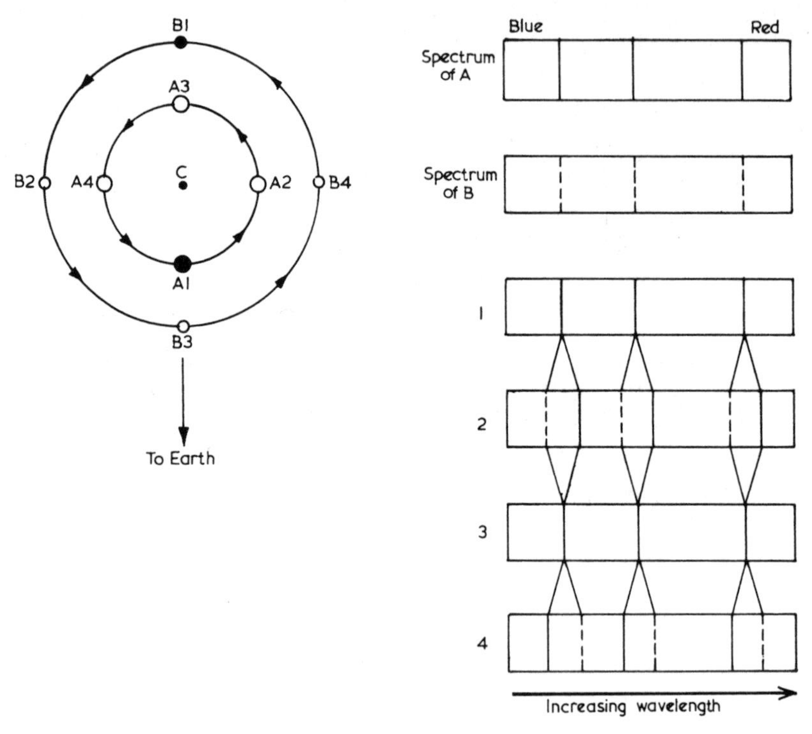

Fig. 27 *Spectroscopic binary.* Two stars, A, and B, revolve around their common centre of mass, C; their orbits are such that we see them roughly "edge on". Each star produces a spectrum consisting of a rainbow band of all colours, superimposed upon which is a characteristic pattern of dark lines; for simplicity, the stars are shown as having identical spectra. When the stars are in position 1 they are crossing our line of sight and, at that instant, neither approaching nor receding from Earth. The spectral lines of both stars are neither red-shifted nor blue-shifted, and through the spectroscope we see only one set of spectral lines (the two sets of lines are superimposed). At position 2, A is receding and B is approaching, with the result that A's spectral lines are red-shifted and B's spectral lines are blue shifted. Accordingly, the two sets of spectral lines are separated in wavelength, and they show up separately when seen through the spectroscope. At position 3, the stars are again crossing the line of sight while, at position 4, A is approaching and B is receding: the resultant effects on the observed spectrum are as shown. As the stars revolve around each other the pattern of spectral lines changes in a periodic fashion, so revealing the presence of two stars.

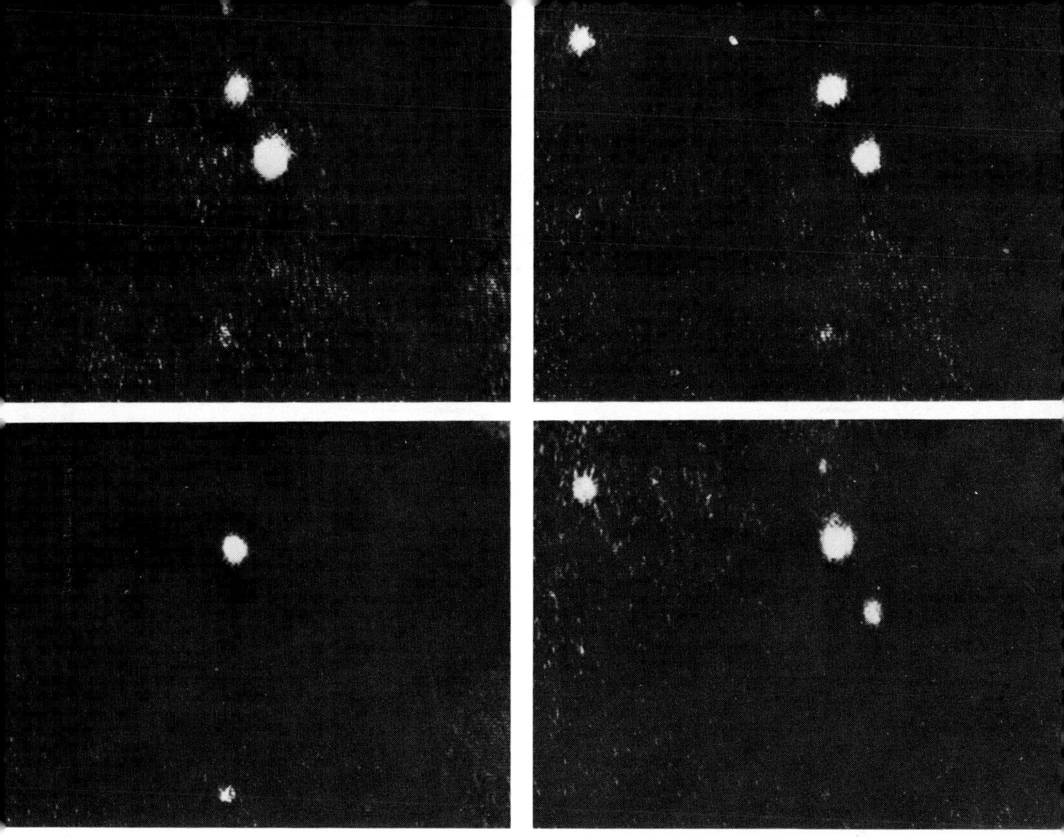

Four different phases of the pulsar NP 0532 in the Crab nebula. The pulsar, which flashes on and off with a period of 0.033 seconds, is seen in the 'off' state in the frame on the lower left.

Cygnus X-1. An X-ray photograph taken by the Einstein Observatory showing the strong X-ray source associated with a binary system which contains the best-authenticated candidate for identification as a black hole. The most popular model suggests that the X-rays are emitted from material in the accretion disc surrounding a black hole of between about 8 and 11 solar masses. However, the evidence in favour of this suggestion is not conclusive. (Courtesy R. Giacconi, High Energy Astrophysics Division; Harvard/Smithsonian Center for Astrophysics.)

The High Energy Astronomical Observatories (HEAO 1, lower left; HEAO 2, lower right; HEAO 3, upper centre) constructed for NASA by the TRW Defense and Space Systems Group. HEAO 1 and HEAO 2 (known as the 'Einstein Observatory') were launched in 1977 and 1978 respectively to study X-rays from a wide variety of sources, such as clusters of galaxies, quasars, pulsars, and black holes. HEAO 3, launched in 1979 is studying cosmic rays and gamma rays. (Courtesy TRW Inc.)

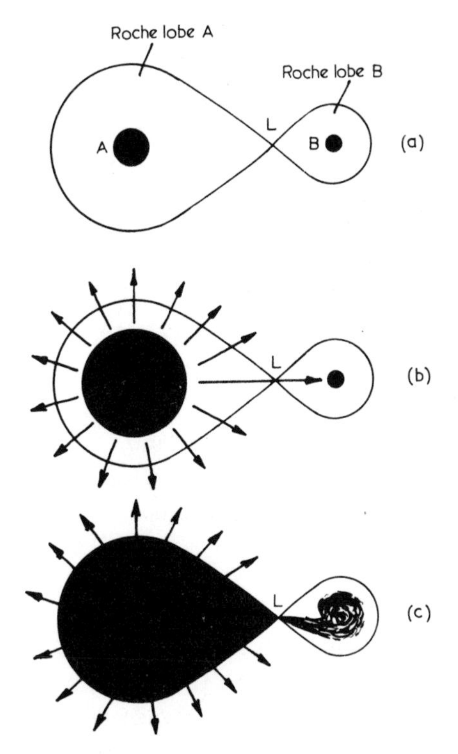

Fig. 28 *Mass transfer in a binary system.* The "figure of eight" shape denotes the equipotential surface drawn around the binary; a particle at any point on this surface experiences the same value of the gravitational field (star A is more massive than star B). (*a*) Both stars lie well within their Roche lobes, and retain their matter. (*b*) If A should expand almost to fill its lobe, particles from A can readily escape in the form of a stellar "wind"; particles which pass through the cross-over point, L (the inner Lagrangian point), can be captured by B. (*c*) If A should fill its Roche lobe then very considerable mass loss can occur: material which crosses the lobe will be lost into space, while material passing through L will be accreted by B.

If we draw spherical shells in space around a single star, at every point on a given shell the value of the star's gravitational field will be the same. If we perform the same exercise for a binary system the shapes of the surfaces (*equipotential* surfaces) over which the gravitational field has the same value will be more complicated, but there is one "figure of eight" equipotential which is of particular importance (fig. 28). The two sections of the curve, each containing a star, are known as *Roche lobes*, after the French mathematician, E. Roche (1820–1883), who investigated gravitational tidal effects in the nineteenth century.

Matter contained inside a Roche lobe remains under the gravitational influence of the star in that lobe, and cannot escape; but matter which crosses beyond the figure of eight will escape into space. As we shall see, in the later stages of their lives most stars expand and, if a star expands beyond its own Roche lobe, it can lose a great deal of its material into space. Some of its matter will pass through the cross-over point of the "8" (the inner Lagrangian point) to be captured by the companion star. In this way an expanding star can "dump" vast quantities of matter onto its companion, whose evolution is, needless to say, seriously affected by it. Considerable mass loss can occur even if a star does not quite overflow its Roche lobe, as material ejected from the star in the form of "stellar wind" can escape or be captured by the companion.

Mass transfer in binary systems is of the utmost importance in a wide variety of astrophysical situations.

Stars and their life cycles

Even when viewed through the largest optical telescopes, stars still look just like points of light. At present the Sun is the only star whose surface we can examine in detail; for the rest, astronomers have to build up their understanding of stars on the basis of a limited range of observations of properties like brightness, colour, motion and so on. As we have seen, the masses of stars can be determined by analysing the motion of binary systems, and it turns out that most stars have a mass which lies between about 0.1 and 10 solar masses, although there are some stars as massive as 50 or even 100 Suns. Another important quantity is *luminosity*, the amount of radiant energy of all kinds emitted by a star in every second, i.e., the star's power output. The luminosity of the Sun, which we often use as a unit when talking of other stars, is about 4×10^{26} watts. The luminosities of stars range from less than one ten-thousandth to more than 100,000 solar luminosities.

The colour of a star gives an indication of its temperature, red stars being relatively cool, with surface temperatures of around 3000 K, yellow stars like the Sun being hotter (about 6000 K), and blue stars very much hotter (20,000–30,000 K). Much more detailed information about a star can be obtained by analysing its spectrum. The hot dense interior of a star emits a continuous spectrum, a rainbow band of all wavelengths of light, whereas in the more rarefied stellar atmosphere atoms absorb light at particular wavelengths (see Chapter 4), giving rise to characteristic patterns of dark absorption lines against the bright background of the continuous spectrum. The lines which are present, and the appearance of those lines (e.g., thin, thick, faint, conspicuous) depend upon a variety of factors relating to the star—temperature, chemical composition, density, rotation rate, magnetic field strength and so on—and analysis of spectra provides a large proportion of our knowledge of the physical natures of stars.

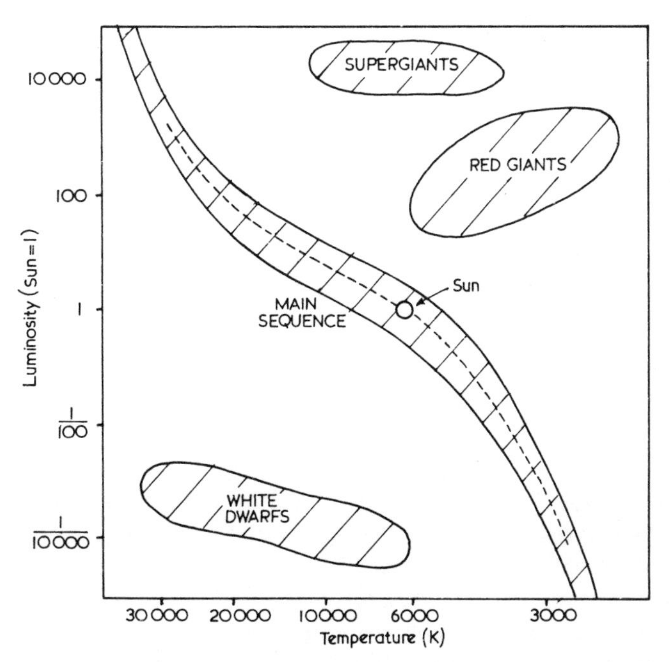

Fig. 29 *The Hertzsprung–Russell diagram.* Stars are assigned positions on the diagram according to their luminosity and their temperature. The Sun, for example, we take to have luminosity = 1, and it has a surface temperature of just under 6000 K. Red giants are cool but highly luminous stars; white dwarfs are hot but of very low luminosity. A typical star will during its lifetime pass onto the main sequence, and there spend the greatest part of its life; it will then become, successively, a red giant and a white dwarf. Some stars have rather different destinies, as discussed in the text.

If stars are plotted according to their luminosities and their temperatures on a diagram (fig. 29), known as the Hertzsprung–Russell diagram after the two astronomers who in the early twentieth century independently noticed the relationship, it is found that the majority of stars lie in a band called the *main sequence* which runs from the top left (hot, bright stars) to the bottom right (cool, faint stars). This tells us that, as a general rule, the hotter the star the more luminous it is, a fact which does not seem particularly surprising.

However, there are stars which do not fit this pattern. In particular, in the upper right of the diagram there are some cool orange and red stars which are highly luminous. Since stars of the same temperature emit roughly the same amount of light from each square metre of their surfaces, the high luminosities of these stars imply that they must be very much larger than main-sequence stars of the same temperature. Consequently, these stars are known as *red giants* and *supergiants*. The conspicuous star Betelgeuse in the constellation Orion is a red giant so large that if it were placed where the Sun is the orbit of the planet Mars would be easily contained inside it.

A significant number of stars which have high surface temperatures nevertheless have very low luminosities. Applying the same kind of argument as above, we conclude that these stars must be very small—hence the name, *white dwarfs*. A typical white dwarf would be about the size of the Earth.

Stars like the Sun live for billions of years, and it is clearly impossible for astronomers to follow an individual star through its life cycle from its birth to its death. Instead, by looking at a sufficiently large number of stars, we can see stars of all sorts of different stages in their life cycles, and by examining these different stages we can piece together the different evolutionary steps through which an individual star passes in its lifetime.

Stars are born inside clouds of gas. We can see some of these clouds directly, those which emit light owing to the presence within them of highly luminous young stars; they are known as *emission nebulae*. If a localized region of such a cloud exceeds a certain mass, the net gravitational attraction of all the atoms in that region causes it to begin to fall together. As the cloud shrinks, so its temperature rises. While the cloud was large and distended it possessed a great deal of potential energy—just as a weight suspended above the surface of the Earth has a quantity of potential energy, equal to its mass multiplied by its height above the surface and the acceleration due to gravity. As the particles in the cloud fall together this potential energy becomes converted into kinetic energy, or energy of motion—just as, if we release the suspended weight, it falls with ever-increasing speed until it hits the ground. Part of the energy of the particles contributes to raising the temperature of the cloud, while the remainder is lost into space in the form of radiation. The collapsing cloud becomes a *protostar.*

In fact, there are many problems associated with our ideas of star formation. Because of the internal temperature and pressure of typical interstellar clouds, it is difficult to see how a mass of less than about ten thousand solar masses (several hundred thousand solar masses if magnetic fields are taken into account) can collapse under its own weight unless acted upon by some kind of "trigger". The trigger could be the encounter of the cloud with a denser region—for example, the cloud might drift into one of the spiral arms of our Galaxy—or perhaps the "shockwave" from the explosion of a supernova. There is good reason to suppose that the collapse of a large cloud will result in its fragmenting into a large number of smaller clouds which go on to form individual stars. Whatever the mechanism involved, we know that young stars *are* found in clouds.

As the temperature and density of the protostar build up, the rate of contraction slows down. Eventually the temperature and density in the core become sufficiently great for thermonuclear reactions to begin, fusing together nuclei of hydrogen to form nuclei of helium. In this process about 0.7% of the matter involved is destroyed, being converted into energy in accordance with Einstein's formula, $E=mc^2$: in the Sun, for example, about 4 million tonnes of matter is converted into energy every second. When this process becomes established, a state of balance is achieved whereby the outward-acting pressure of the hot gas inside the star counteracts the gravitational self-attraction which is endeavouring to compress the star. It becomes a stable main-sequence star with a temperature and luminosity determined by the mass of the star—the more massive the star, the more luminous it is.

A star remains on the main sequence for the major part of its life. The more massive it is, the shorter its main-sequence phase: the Sun, for example, is expected to have a main-sequence lifetime of between 10 and 11 billion years (it is now about 5 billion years old), but a star of, say, 10 solar masses, although it has 10 times as much fuel, is thousands of times more luminous, and so squanders its reserves much more rapidly.

Eventually the powerhouse in the central core of the star runs out of hydrogen fuel and shrinks under the weight of the rest of the star. As it does so, the temperature rises in the surrounding shell of hydrogen to an extent sufficient to allow the hydrogen-to-helium fusion reaction to take place progressively further and further out from the original core. The net result is an increase in the star's luminosity, and the star expands until the outward-acting pressure and the inward-acting gravitational force are once more in equilibrium. By this time a new fusion reaction, converting helium into carbon, has started up in the core and the star reaches another fairly stable state—it becomes a red giant.

A star does not remain a red giant for long, for it is using up its remaining reserves rapidly. When all possible nuclear fuels have been used up (in massive stars these reactions can continue until iron is produced in the core) the star can no longer support itself, and gravity wins the next round of the contest.

Stars having a final mass less than about $1\frac{1}{4}$ solar masses are compressed by gravity until they reach densities of about 10^8–10^9kg per cubic metre. They become white dwarfs, stars which, over aeons of time, cool down eventually to become dark black dwarfs. The material in a typical white dwarf is so densely compressed that a teaspoonful of it, if brought back to Earth, would weigh several tonnes.

A star which at the end of its life has a mass above this limit cannot become a white dwarf. Instead, it may become a *neutron star*, a star so severely squashed that its protons and electrons have been forced to combine to form electrically neutral neutrons. A typical neutron star would have a radius of only about 10km, and the density of its material would be of the order of 10^{18}kg per cubic metre; i.e., on Earth a teaspoonful of it would weigh several billion tonnes!

It is thought that many—if not all—neutron stars are formed in the following way. When the cores of stars many times the mass of the Sun cease to be able to support themselves against gravity they collapse, the outer layers of the star falling in on top. The tremendous quantity of energy released results in a supernova explosion, which blows most of the outer body of the star into space. The best known example of such an event was the "guest star" seen by Chinese observers in the year 1054, a star which flared into view, becoming bright enough to be seen in daylight, before fading from view after

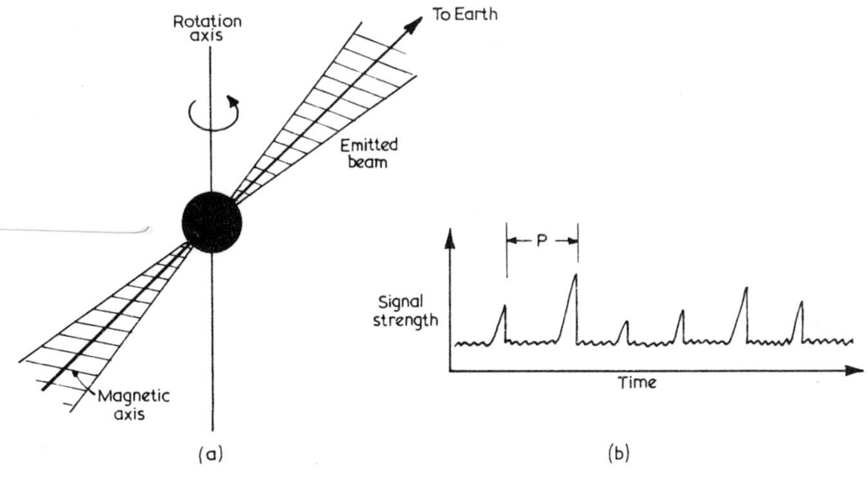

Fig. 30 *A pulsar.* A pulsar is believed to consist of a rapidly rotating neutron star (*a*) from which radiation is emitted in two oppositely directed narrow beams. When the beam is directed towards the Earth, we receive a pulse of signal, like a flash from a lighthouse, the resulting pattern of signals being shown in (*b*). The interval between pulses, the period, P, is remarkably constant and, for known pulsars, has values between about 0.03 seconds and a few seconds. Over the centuries, P slowly increases, so measuring it can give some indication of the relative age of a pulsar. It should be stressed that there are several theoretical models of how, and from where, the pulsar emission is released. This illustration shows one model, where the emission comes from the region of the magnetic poles; in another, emission comes from above the equator.

a few months. We can still see what is called the Crab Nebula, the turbulent remnants of that exploding star.

At the heart of the Crab Nebula lies a strange flashing source of radio waves, light and X-rays, emitting a pulse of radiation some thirty times per second. It is known as a *pulsar*. Well over three hundred pulsars are known (rather too many, it appears, to fit in with the expected number of supernova events in the Galaxy); the first was discovered by Jocelyn Bell (now Jocelyn Burnell) and Antony Hewish during a radio survey carried out at Cambridge in 1967. It is generally accepted that a pulsar is a rapidly rotating neutron star from which radiation is emitted in the form of narrow beams (fig. 30). Each time the neutron star spins round our way we pick up a pulse—in the same way as we might see pulses of light from a lighthouse.

As was first pointed out in 1939 by J. R. Oppenheimer and H. Snyder of the University of California, there is a maximum possible mass a neutron star of moderate rotation rate can have. Although the value of this maximum is not yet well established, it is believed to be not more than about 2 or 3 solar masses. There are many ways in which stars can lose mass towards the end of their lives—stellar wind, mass transfer in binary systems, supernovae and so on—but there are significant numbers of stars 10, 20 or even 50 times the mass of the Sun and it seems highly improbable that *all* of these will lose sufficient mass to come below the limit. Theory suggests that, once a star or stellar core of mass in excess of that limit begins to collapse under its own gravitational attraction, *nothing* can halt the collapse. The matter comprising the star collapses without limit until, in principle, it is all squashed into a point. As the collapse progresses, the force of gravity at the surface rises until a stage is reached when not even light can escape. The star disappears, forming what has come to be known as a *black hole*.

Black holes represent regions in space of extreme gravitational fields—regions which require for their treatment the full armoury of General Relativity. It is the possibility of detecting these bizarre entities which has, more than anything else, triggered the great upsurge of interest in gravitational physics which we see today.

7
Extremes of Gravity

To the theoretician, black holes are exciting because they represent a frontier at which conventional theories of gravitation are stretched to the limit, where new refinements or revolutionary theories become necessary. To the astrophysicist, they are of the utmost interest in that they may provide explanations for some of the most perplexing phenomena in the Universe. To most of us they have a compelling fascination because of the weird effects on space and time with which they are associated. To the science-fiction writer, they are a mixed blessing: on the one hand they open up fascinating possibilities, but on the other they represent an area where truth is stranger than fiction, where fiction writers can be upstaged by the bizarre predictions of sober scientists working with well established laws of nature.

What is a black hole? Essentially it is a region of space into which matter has fallen and from which nothing can escape; within a black hole, gravity is so powerful that not even light can move outwards. The term "black" is highly appropriate, for if no light can emerge from within a black hole it must be a truly black entity; the term "hole" is also appropriate, as unlimited quantities of matter or radiation can fall in, and in that sense it is a bottomless pit. As we shall see in later chapters, black holes are not nearly as simple as this, but the basic concept remains as given above.

The term "black hole" was coined in 1968 by Professor J. A. Wheeler of Princeton University, but the idea that such entities might exist goes back much further. In a sense, the idea goes back about two hundred years. In a paper read to the Royal Society of London in 1783, and published in their *Philosophical Transactions* the following year, the English physicist John Michell (1724–1793) alluded to the possibility that if, as Newton had proposed, light consisted of a stream of particles, then these particles should be affected by gravity in the same way as were material bodies. Therefore, he suggested, light moving away from a massive body should be slowed down. In particular, he pointed out, light could not escape from a body having the same density as the Sun but a 500 times greater radius, since the escape velocity* at the surface of such a body would be greater than the speed of light.

Some thirteen years later, the great French mathematician Pierre Simon Laplace (1749–1827) made similar remarks in his book *Exposition du Système du Monde*, suggesting that a body with 250 times the radius of the Sun and the same density as the Earth would be invisible because its light could not escape. Bearing in mind that the density of the Earth is about four times greater than the density of the Sun, the suggestions made by Laplace and by Michell are in quite close agreement.

It is a straightforward matter to apply this concept to any body of mass M to find the radius within which it would have to be compressed in order that its escape velocity be greater than the speed of light. If we set the velocity of escape (v_e) equal to the speed of light (c), we can write the formula $c=\sqrt{2GM/R}$. Manipulating this, we find that a body of mass M has an escape velocity greater than the speed of light if it is compressed within a radius R where $R=2GM/c^2$.

The possibility that there might exist objects of the type hinted at by Michell and Laplace aroused little interest at the time, and was allowed to languish for well over a century. However, early in 1916, a few months after Einstein's General Theory of Relativity was published, the German theoretician Karl Schwarzschild (1873–1916) solved the field equations describing space-time in the vicinity of a spherical lump of matter. His solution can be interpreted in the following way. If a mass M is compressed within a certain radius

*The escape velocity at the surface of a body is the minimum speed at which a projectile must be thrown upwards in order that it may continue to move away forever without falling back down again. When an object is thrown with a given velocity, it will reach a particular height before falling back to the ground. If it is thrown at a higher velocity it will reach a greater height before falling back, and if it is thrown with *precisely* the escape velocity it will slow down as it moves away but its speed will not drop to zero until it attains an infinite distance. If a body is projected at a speed in excess of escape velocity, then its speed will never drop to zero. The escape velocity (v_e) is readily calculated from the formula $v_e=\sqrt{2GM/R^2}$, *where* G is the gravitational constant, M the mass of the body from which escape is to take place, and R the distance from the centre of that body. Escape velocity at the surface of the Earth is 11.2km per second.

(which we shall call R_s to denote *Schwarzschild radius*) space-time is distorted so severely that light is unable to escape. Since nothing is permitted to travel faster than light, no material object or signal of any kind can escape from within that radius. A ray of light emitted radially outwards *at* R_s would remain there, hovering, for the rest of time; in effect it would be running flat out to stand still. The region of space out of which nothing can escape is what we today call a black hole.

From the Schwarzschild solution it turns out that, for any body of mass M, $R_s = 2GM/c^2$. The remarkable thing about this formula is that it is precisely the same as the one derived from Newton's theory of gravitation and the idea of escape velocity. Of course, we should not be amazed that General Relativity and Newtonian theory often give the same answers, for it is only in extreme circumstances that we expect differences between the two theories to be manifest. Nevertheless, a black hole hardly constitutes "normal circumstances", and it is quite surprising to find both theories apparently giving the same answers in this context.

In fact the "black hole" of Newtonian theory is not really the same kind of animal as the one derived from General Relativity. Although it gives a good feel for what is going on, the escape-velocity argument falls down in one respect. You or I cannot possibly throw a ball faster than the Earth's escape velocity, but we can throw it up in the air; the faster we throw it, the higher it goes before falling back. Likewise, on the Newtonian basis, light particles should be able to travel a certain distance away from a black hole before falling back again, even if the escape velocity at the surface were considerably greater than the speed of light. With a Schwarzschild black hole, on the other hand, light emitted at the "surface"—i.e., at the Schwarzschild radius—cannot move outwards at all. The agreement between Newtonian and Einsteinian theory is rather superficial.

The Schwarzschild black hole

A black hole will form when a given quantity of matter is squeezed inside its Schwarzschild radius. As we have seen in Chapter 6, one way in which this may happen is when a massive star at the end of its life collapses under its own gravitational attraction. If when collapse begins the final mass of the star, its core or remnant, exceeds

3 solar masses (the limit may be less than this) then so far as we can tell at present there is no force in nature which can halt the indefinite collapse of that star: the collapse will continue until all of the star's material is compressed into a point known as a *singularity*. In the singularity, matter is infinitely compressed to infinite density by infinitely powerful gravitational forces; putting it another way, the space-time curvature is infinitely great at the singularity. Infinite forces and densities are not phenomena with which present-day physics can cope: the laws of nature as we understand them break down at a singularity. As for the matter which made up the star, it appears that it is crushed out of existence at the singularity.

It has been argued that perhaps some new force comes into play to prevent the ultimate collapse to truly infinite density. For the present this is a matter of pure speculation. Gravity is such a weak force, compared to the other forces of nature, that quantum effects—which are readily apparent with the other forces—are not expected to become dominant with gravity until one is dealing with very short distances indeed, possibly as short as the Planck length, 10^{-35} metres. As yet we do not have a workable quantum theory of gravity. Perhaps when we do, we will find that true singularities do not occur in nature, but on the present evidence it looks as if General Relativity will hold good inside a black hole all the way down to the immediate vicinity of the singularity, and that matter *will* be compressed within a microscopic volume at the centre of a black hole.

As the collapsing star passes through its Schwarzschild radius it disappears from view, because light from its surface is no longer able to reach us. We say that a *horizon* has been formed: we cannot see what is happening inside this horizon. We have every reason to suppose that the star continues to collapse into a singularity, but there is no way in which we can see this happen, and there is no way in which we can obtain information about the fate of its constituent material. The black hole resulting from the collapse of the star (fig. 31) is the spherical volume of space centred on the singularity, and with radius equal to the Schwarzschild radius. The boundary of the black hole is known as the *event horizon* because no knowledge of any events which take place inside that horizon can be communicated to the outside Universe.

A black hole has, of course, no solid surface. If you were to cross the event horizon you would not notice that space there was different

Table 2 Examples of Schwarzschild radii

Object	Mass	Schwarzschild radius (R_s)*	Density of collapsing mass as it reaches its Schwarzschild radius (kg/m^3)*
small mountain	10^{12}kg	10^{-15}m	10^{56}
small asteroid	10^{18}kg	10^{-9}m	10^{44}
The Earth	6×10^{24}kg	1cm	10^{30}
The Sun	2×10^{30}kg $=1M_\odot$†	3km	10^{19}
Massive star	$10M_\odot$	30km	10^{17}
Possible collapsed mass in active galactic nucleus	$10^8M_\odot$	3×10^8km	10^3 (water density)
Entire Galaxy	$10^{11}M_\odot$	0.03 light-years	10^{-3}

*Values of R_s and density are approximate, being rounded off to the nearest power of 10 in the latter case.
†$1M_\odot$ is the mass of our Sun.

from space anywhere else, but once within that boundary you would find yourself unable to move outwards, and would fall inexorably into the central singularity. The event horizon is a one-way boundary. Material objects, light, or any other kind of radiation can fall into a black hole, but nothing can come out.*

In principle, if not in practice, almost any amount of material can

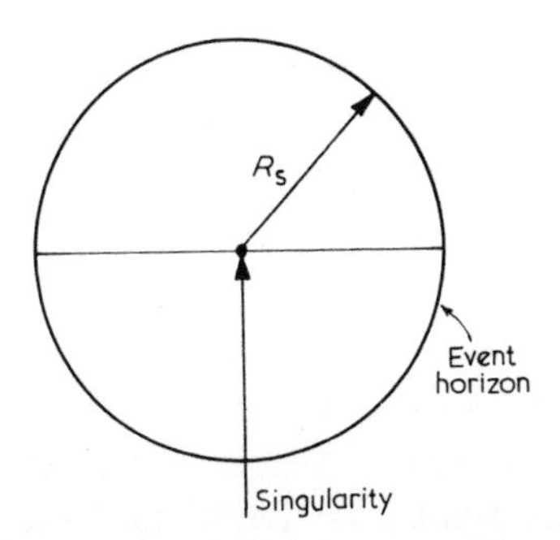

Fig. 31 *A Schwarzschild black hole.* A simple, nonrotating black hole is a spherical region whose radius equals the Schwarzschild radius (R_s) corresponding to the mass of the hole. The boundary of the hole is called the event horizon; no signal, ray of light or material body can escape from within this boundary. At the centre, material is crushed by infinite gravitational forces; i.e., there is a central singularity.

* In later chapters we shall see that the situation is not quite so clear-cut as this.

be made into a black hole. For every mass, there is a Schwarzschild radius within which it can be compressed (examples are given in Table 2). To give some idea of scale, the Schwarzschild radius for the Sun is just under 3km, and so it would be necessary to compress the Sun within that radius to convert it into a black hole. As the present radius of the Sun is about 700,000km its mean density as it reached its Schwarzschild radius would be over ten million billion (10^{16}) times greater than the density of water. If, in a fit of megalomania, the archetypal mad scientist wished to convert the Earth into a black hole, the Earth would have to be squeezed within a radius of less than one *centimetre*!

There is no natural way that either the Earth or the Sun could be made into a black hole in the present state of the Universe. As we have seen, stars which end their lives with masses less than 2 or 3 solar masses are likely to end up as white dwarfs or neutron stars. However, there are many stars which are well above this limit and, although there are many ways in which they can shed excess mass towards the end of their lives, nevertheless it is highly likely that some of these high-mass stars *will* collapse to form black holes.

A star of 10 solar masses has a Schwarzschild radius of nearly 30km. Since the volume of a sphere depends on the cube of its radius, and the radius of a black hole depends on its mass, the density of matter at the instant when a star reaches its own Schwarzschild radius is less for higher mass stars than it is for lower-mass stars. Thus a star of 10 solar masses will have when it reaches that radius an average density of only (!) 10^{14} times that of water (10^{17}kg per cubic metre). In the previous chapter we saw that neutron stars are likely to have mean densities of around 10^{18}kg per cubic metre. Since the evidence for the existence of neutron stars is very good indeed, it is clear that matter *can* be compressed to these enormous densities and, as we see here, a collapsing star of 10 solar masses would be less dense than a neutron star at the instant at which it formed a black hole. Of course, the collapse within the black hole would continue until infinite density was attained, but the fact remains that black holes can form around matter having densities *less* than those which are already believed to exist in known objects.

Taking this question a little further (Table 2) we see, for example, that a black hole containing a hundred million (10^8) solar masses would have a radius of about 300 million kilometres (twice the radius of the Earth's orbit), and the average density of the collapsing material just as the event horizon formed would be about the same as that of water. A black hole of a few billion solar masses would form at a density comparable to that of the air which we are breathing. It must be emphasized that, once a given mass is compressed within its Schwarzschild radius, nothing can prevent its subsequent indefinite collapse; but to *form* the black hole in the first place incredible compression is not necessarily required.

In the Universe we may find black holes of masses lying between 2 or 3 and about 100 solar masses resulting from stellar collapses. Supermassive black holes, containing thousands, millions or billions of solar masses may also exist and could be forming, or growing, in the Universe today. It has also been suggested that, if the Universe began in a hot dense Big Bang, conditions in the earliest moments may have been such that even quite small amounts of material could have been squeezed sufficiently to form "mini black holes". The mass of a mountain could be contained within a black hole about the size of an atomic nucleus, and it may be that such objects exist. We shall have more to say about maxi and mini holes in later chapters, but at this point it is sufficient to note that black holes may span a colossal range of size and mass.

Near the event horizon

In the vicinity of a non-rotating Schwarzschild black hole, a number of curious effects are to be expected. Let us investigate a 10-solar-mass black hole with the aid of a hypothetical experiment (a "*gedanken*-experiment", or "thought-experiment") whereby a volunteer astronaut—in the interests of science, or perhaps to avoid paying income tax arrears—agrees to plunge into the black hole, sending back information for as long as he can. He realizes that this is a one-way mission: once he has crossed the event horizon there will be no way back. We—more wisely—will observe events from a safe distance, well beyond the event horizon.

One rather unpleasant effect to which the infalling astronaut would be subjected would be tidal forces. As we saw in Chapter 3, tidal forces arise as a result of the difference in gravitational attraction exerted on different parts of an extended body. Standing on the surface of the Earth (even ignoring the effects of the Sun and the Moon) we are subject to tidal effects due to the Earth alone. If you are standing upright, your feet are closer to the centre of the Earth than is your head, and for that reason are subject to a stronger attraction. Of course, the effects are infinitesimal and we are neither aware of them nor even able to measure them in everyday circumstances.

Close to the event horizon of our black hole things are very different. 10 solar masses of material have been compressed within a radius of 30 kilometres, and the gravitational force rises very sharply as we approach the event horizon: close to it, the infalling astronaut would be subject to a tidal force roughly equal to that experienced by someone hanging from a bridge with the entire population of London or New York swinging from his ankles!

The infalling astronaut would be stretched and torn apart by a "cosmic rack" of ever-increasing severity long before he reached the event horizon. Once through it, his shattered remains would plunge into the central singularity to be crushed out of existence. At least the end would be mercifully swift. Falling virtually at the speed of light, the remains would reach the singularity about one ten-thousandth of a second after crossing the horizon.

The tidal forces experienced at the event horizon are markedly less for black holes with greater mass. In fact, the tidal forces at the event horizon are inversely proportional to the square of the mass, so that the tidal stress experienced entering a 20-solar-mass hole would be only one quarter of that encountered in falling into a 10-solar-mass hole. If we assume, for the sake of argument, that the human body can for a brief period withstand being stretched by a force about ten times greater than body weight, then an astronaut should be able to penetrate the event horizon of a 10,000-solar-mass hole without being torn apart. At the event horizon of a 100-million-solar-mass black hole, the tidal effects would be less than those experienced here on Earth due to the Earth alone; it would be possible to stumble into such a hole without noticing the fact, although it would then be too late to avoid eventual destruction.

If the infalling astronaut is equipped with a transmitter which emits a pulse or signal at regular intervals according to the clock he carries with him (his proper clock), then from his point of view the intervals between transmitted pulses remains precisely constant. The situation appears rather different to us. Initially the time intervals between arriving signals remain constant, and equal to those be-

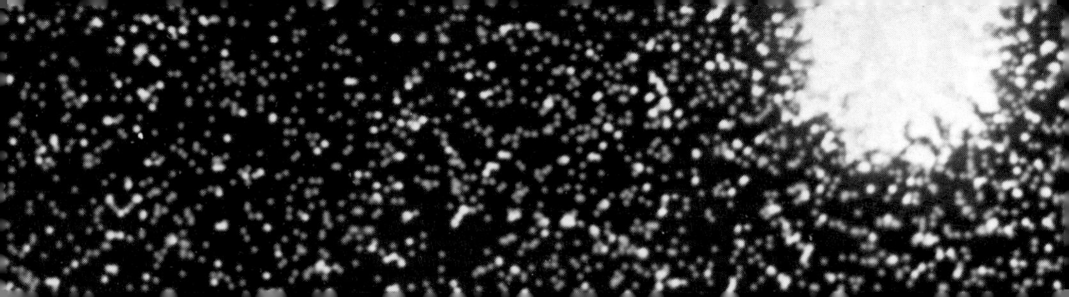

Cygnus A. The brightest radio source in the sky, Cygnus A coincides in position with the peculiar galaxy shown here, a galaxy located at a distance in excess of a billion light-years. At one time it was thought that this object represented two galaxies in collision, but it is now known to be a single galaxy having two giant radio-emitting lobes. (Photograph from the Palomar Observatory, California Institute of Technology.)

tween transmissions according to the astronaut's timescale. As he approaches the event horizon, discrepancies become increasingly more obvious: the intervals between successive pulses according to our timescale become progressively longer and longer. From our point of view, the astronaut's clock is slowing down due to the effects of gravitational time dilation. The deeper he penetrates into the gravitational field of the hole the more obvious the effect becomes (fig. 32) until, when he reaches the horizon itself, we would conclude that his clock has stopped altogether. By our reckoning the astronaut takes an infinite time to cross the event horizon, and we may well imagine that, with a sufficiently powerful telescope, we should be able to see a frozen image of the shredded volunteer, hovering on the brink of the black hole for the rest of eternity.

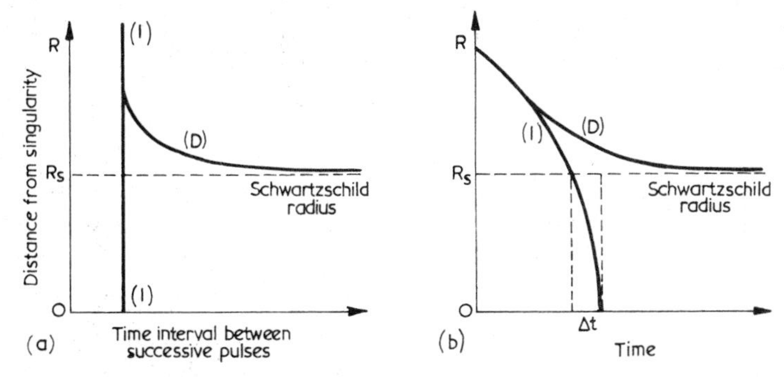

Fig. 32 *Contrasting views of time near a black hole.* (*a*) An astronaut falling into a black hole signals to a distant observer by emitting flashes of light at intervals which are precisely regular according to his (the astronaut's) proper time (I). According to a distant observer (D), the time interval between the received flashes becomes longer and longer as the astronaut approaches the event horizon. (*b*) According to the astronaut's proper time (I), he continues to accelerate as he approaches the hole, and, after crossing the event horizon, reaches the central singularity in a very short period of time (Δt; for a typical stellar-mass black hole, Δt might be about one ten-thousandth of a second). According to D the astronaut never gets there, but takes an infinite time to cross the event horizon.

The astronaut sees things very differently. According to his clock—and every other means of local time measurement, atomic clocks, biological clocks or whatever—time is flowing in the usual way, at a uniform rate. He crosses the event horizon and plunges into the central singularity in a tiny fraction of a second; and that is a very real, painful and final event for him. Yet *we* conclude that he has never crossed the event horizon at all.

Who is correct? Both, or neither, according to your point of view. We are correct in the observations made in our own frame of reference; likewise the infalling astronaut is correct in the deductions made in *his* frame of reference. There is no absolute standard of time in the Universe: both observers are entitled—with equal validity—to their points of view.

But surely the same line of argument could be applied to the collapsing star which went to form the black hole in the first place? If we were to watch a collapsing star we should see the collapse slow down and halt just at the Schwarzschild radius, and a frozen image of the star should remain in view forever more (a "nearly black hole"). In practice this would not be so. Another relativistic effect, intimately related to the gravitational time dilation, also comes into play—the gravitational red-shift. Indeed, the gravitational red-shift and time dilation are part and parcel of the same phenomenon. Close to the event horizon the red-shift becomes very powerful indeed. As the collapsing star approaches closer and closer to its Schwarzschild radius, the wavelengths of radiation emitted from its surface become progressively more and more red-shifted into the longest-wave part of the spectrum, becoming progressively weaker as they do so. The frequency of the radiation declines rapidly until the time interval between the penultimate and the final wavecrest to reach the distant observer becomes infinitely long. As a result, the collapsing star (or the infalling astronaut) vanishes from view within a tiny fraction of a second as the event horizon is reached.

If another astronaut is despatched to check on whether or not the first one really has fallen through the event horizon, he will never catch up with him. The first astronaut has fallen in ahead of him, and the second one, too, according to his own proper time, plunges in, very rapidly, to meet the same fate as the first. In practice, then, black holes *will* form, collapsing stars *will* vanish from sight, and infalling astronauts *will* disappear into them. No observations, from any frame of reference, can show otherwise.

Inside the black hole

We cannot observe the interior of a black hole from the outside. Although an infalling astronaut could, in principle, make observations of the interior—particularly if he chose to enter a supermassive black hole where he might have hours, or even days, available before being torn apart by tidal forces—he would be unable to communicate this information to us without the aid of (forbidden) faster-than-light signals. Nevertheless, we have great confidence that general relativity can be used to describe the interior of the black hole right down to, although excluding, the singularity itself.*

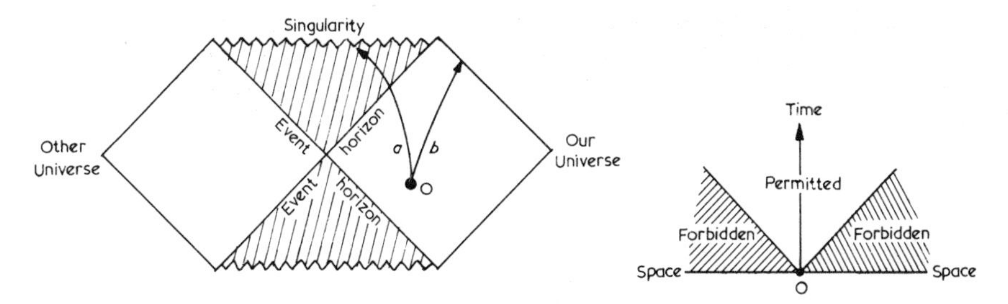

Fig. 33 *Space-time map (Penrose diagram) for a Schwarzschild black hole.* Time is plotted in the vertical direction and space in the horizontal direction (details of the mapping process are given in the text). Our Universe lies on the right-hand side of the diagram, while there appears to be another, inaccessible, universe on the left. Matter particles must follow time-like paths—i.e., lines inclined to the vertical axis by less than 45°—and for this reason it is clear that any particle *a* which crosses the event horizon must strike the singularity; particle *b* remains safely outside the event horizon. (Compare with fig. 35, page 136.)

We encountered the concept of a space-time diagram in Chapter 4 (see page 74). We can use a type of space-time map developed by Professor Roger Penrose of Oxford University, and hence generally referred to as a Penrose diagram, to depict the black hole and the rest of the Universe on a single sheet of paper.

Any map introduces distortions—for example, the familiar Mercator projection used in maps of the Earth represents details close to the equator quite faithfully, but introduces progressively increasing distortions towards the poles. Penrose's technique (known as conformal mapping) produces a space-time map for a Schwarzschild black hole which looks like fig. 33. The whole of space-time outside

*One implication of this is that at no point in his fall towards the singularity would the astronaut exceed the locally measured speed of light. According to Newtonian theory the astronaut would suffer ever-increasing acceleration, striking the singularity at an indefinitely high velocity, far in excess of the speed of light.

the black hole is mapped into the right-hand segment of the diagram, while the lines at the right-hand edge represent infinitely distant regions of space-time stretching from the infinite past (bottom) to the infinite future (top). The event horizon is represented by lines inclined to the vertical (time-direction) by angles of 45°. In the space-time diagrams which we encountered earlier such lines represented the paths of light-rays. Slower-moving particles have world-lines inclined by less than 45° (i.e., they follow time-like trajectories). Since a ray of light directed outwards at the event horizon remains at a constant distance from the singularity, although an infalling observer would still conclude that the ray was travelling at the speed of light (running flat out to stand still), it seems reasonable to denote the horizon by lines inclined in this way. The singularity maps into a horizontal line across the top of the diagram; the singularity is space-like.

A particle falling into a black hole must follow a time-like path, for nothing is permitted to exceed the speed of light. Looking at the map, it is clear that nothing which falls into a Schwarzschild black hole can avoid falling into the central singularity, since even inside the black hole particles must pursue paths inclined to the vertical by less than 45°. There is a fundamental change in the character of space-time which occurs when the event horizon is crossed. Whereas in the outside Universe objects are free to move around in any direction, inside the event horizon, the only possible type of motion leads to destruction at the singularity.

The complete Penrose diagram is symmetrical and shows the apparent existence of another universe on the "other side" of the black hole. The solutions of the equations describing space-time in the vicinity of a Schwarzschild black hole have a symmetry which suggests that a black hole connects our Universe with another one. Extending the rubber-sheet analogy (page 92), the indentation in the rubber sheet representing a black hole in our Universe eventually opens out into another rubber sheet representing another asymptotically flat space-time—another universe.

Does this "other universe" have any physical significance, or is it merely a mathematical curiosity arising from the nature of the solutions to the equations?* The question is entirely hypothetical in the case of a Schwarzschild black hole, because there is no way of reaching the other universe, or of sending a signal into it or receiving one from it. Anything entering the black hole falls into the singularity. In order to make a trip into the other universe, or for something to enter our Universe from the other, a prohibited faster-than-light trip would be involved. The concept of the other universe would thus appear to be no more than a mathematical curiosity. Nevertheless, the hypothetical link between two universes, the Einstein–Rosen bridge—or *wormhole*—has aroused a good deal of interest. It has been speculated that, instead of linking two universes, the bridge might connect different points in our own Universe. Be that as it may, travel from one location in space-time to another through the agency of a Schwarzschild black hole cannot be achieved because of the impossibility of avoiding the singularity.

As for the singularity itself, we can say little about it for our physical laws cannot deal with infinitely compressed matter and infinitely powerful gravity. Penrose and others have shown convincingly that the collapse of a mass of material will always lead to the formation of a singularity and that—at least in the case of spherically symmetrical collapse—an event horizon will always form to hide the singularity from external observers. A singularity represents a region where our laws of nature do not hold good, and a physicist cannot make predictions of what will happen at such a point, or what might emerge from it. If singularities could be seen directly—i.e., if there exist *naked singularities*—then our ability to predict the course of events in the Universe would be undermined; the singularity could always throw an unpredictable spanner in the works. If singularities are always decently cloaked in event horizons, then whatever happens at singularities cannot affect the Universe outside those event horizons. If they are unobservable, then perhaps singularities can just be ignored.

It is less clear whether or not an event horizon must always form round every kind of collapsing body. With the collapse of a spherical mass to form a nonrotating black hole there is no doubt that an

*We frequently encounter mathematical problems which have two solutions, one of which has physical meaning while the other is rejected as being of no significance. The square root of a number is a case in point; for example, the square root of 64 is $+8$ or -8 ($+8 \times +8 = -8 \times -8 = 64$). Consider Kepler's third law: if a planet moves round the Sun at a distance, (a), of 4 astronomical units, the law tells us that the orbital period, (P), in years is given by $P^2 = a^3$. In this case $P^2 = 4^3 = 64$, and the period in years is the square root of 64. We are happy to accept that the period is 8 years rather than -8 years.

event horizon forms, but the collapse of a nonspherical or wildly disturbed body poses problems. It is widely believed, following the work of Penrose, that there exists a *cosmic censorship* principle, which implies that the Universe is constructed in such a way that singularities are *always* hidden behind event horizons, but it has not been proved rigorously for cases other than straightforward, and possibly somewhat idealized, spherical collapse.

Gravity within a black hole dominates all other forces but, if the cosmic censor never slips up, we cannot hope to see the consequences of that domination in the central singularity.

8
Spinning Black Holes

The Schwarzschild black hole which we have been considering so far is physically unrealistic in one crucial sense. If real black holes exist they will have formed from spinning bodies (i.e., bodies with angular momentum) which may even have a net electrical charge. Although most bodies in the Universe are more or less electrically neutral, rotation is a property common to stars, planets and galaxies alike. If a black hole were to form as a result of the collapse of a rotating star, we would expect the black hole to be rotating at a rapid rate; after all, we know that neutron stars spin very rapidly indeed.

As a result of work carried out by B. Carter, W. Israel, D. C. Robinson and S. W. Hawking, it has been shown that black holes, from the point of view of the outside observer, can possess only three distinguishing characteristics. All the properties of a black hole are determined by its mass (M), electrical charge (Q) and angular momentum (J). The reason that these properties are preserved is that they are associated with long-range fields which can exert influence at large distances. The gravitational field associated with M and J diminishes in strength with the square of distance, but extends nevertheless to infinite range; the electromagnetic force associated with Q behaves in similar fashion. When a black hole forms, the gravitational field beyond its event horizon continues to affect the motion of rays and material objects. Likewise, the electromagnetic field associated with any net charge carried by the black hole will influence its surroundings. *No other properties of the matter which went into forming the black hole are preserved.* This aspect of black holes was neatly summarized by Professor J. A. Wheeler in his remark: "A black hole has no hair"; i.e., a collapsing body will settle down quickly to a state characterized only by mass, charge and angular momentum. The black hole has no other distinguishing features—hair, beard, moustache or whatever. Indeed, a black hole formed by 10 solar masses of governmental memoranda would be indistinguishable from one formed by the collapse of a 10-solar-mass star, provided that the initial mass of material had the same degree of rotation and electrical charge. Black holes are rather inscrutable entities.

Solving the field equations for massive collapsed bodies with charge and angular momentum proved to be a difficult task. Between 1916 and 1918 H. Reissner and G. Nordstrøm found solutions for black holes with mass M and charge Q, but it was not until 1963 that the Australian mathematician Roy P. Kerr found a solution for a black hole with mass M and angular momentum J. Only since then has it been possible to discuss realistic black-hole models. Some two years later E. T. Newman and others obtained solutions involving M, J and Q, all the possible characteristics which could be possessed by black holes. The Kerr–Newman solutions appear to cover all possible black holes.

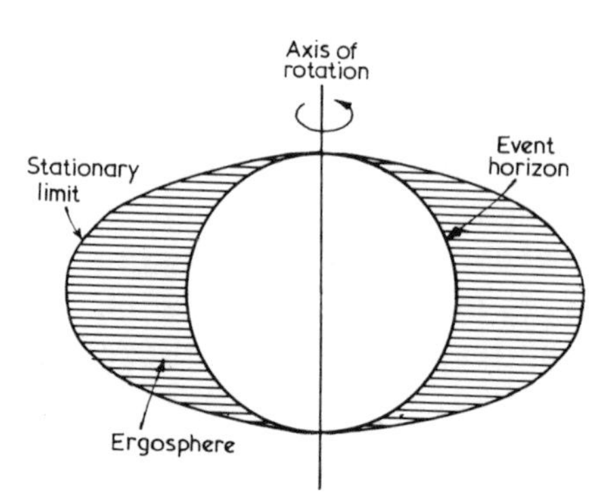

Fig. 34 *A rotating (Kerr) black hole.* Outside the event horizon of a rotating black hole is a region known as the ergosphere. Nothing located within the ergosphere can avoid being dragged round by the rotation of the hole. The boundary of the ergosphere is called the stationary limit, since a particle placed there could just avoid being dragged round if it were capable of travelling at the speed of light. It is possible for a particle to enter the ergosphere and escape, gaining energy in the process (see text).

In point of fact, it seems unlikely that black holes with significant electrical charge can exist—at least for long. If a black hole were formed with, say, a strong net positive charge, it would strongly attract negative charges in its vicinity and would tend to repel positive ones. Over a period of time it would absorb sufficient charges to neutralize its original charge.

Spinning black holes have a number of fascinating properties both inside and outside the event horizon. Outside the event horizon is a region called the *ergosphere* (fig. 34) bounded by a surface known as the *stationary limit* (not, of course, a solid or tangible surface). The stationary limit touches the event horizon at its "poles" and bulges out to the greatest extent above the "equator". Within the stationary limit nothing can stand still: even if you were in a rocket ship capable of attaining nearly the speed of light, you would be unable to avoid being dragged round by the rotation of the hole; it is as if space itself were being dragged round by the hole. In principle, a spacecraft could enter the ergosphere and leave it again by firing its motors in a direction which took it away from the event horizon, but there is no way that it (or a ray of light, for that matter) could remain stationary while inside that region.

The effect is known as the Lense–Thirring "dragging of inertial frames". We can get some idea of why this should happen by thinking back to Mach's principle (see page 59), the idea that every body in the Universe contributes to the property of inertia in every other body. Under normal circumstances it is the very distant matter in the Universe which makes the overwhelming contribution to the property of inertia, but local bodies also must have some effect, however microscopic. A massive rotating body should therefore drag round to some extent the inertial frames in its locality. In principle it should be possible to measure this effect by noting if a Foucault pendulum (a freely swinging pendulum which swings in a "fixed" plane as the Earth rotates beneath it) located over one of the terrestrial poles remains forever swinging in a fixed plane relative to the distant galaxies, or if it very slowly drifts round in the direction of the Earth's rotation. The effect is too small to be measured as yet. However, if all other perturbations could be accounted for, the precession of the orbit of a satellite might become detectable, although the time taken for the orbit to turn round once because of this effect would be about 10 million years.

In 1969 Roger Penrose suggested a way in which energy could be extracted from the ergosphere of a black hole. If a particle possessing a certain amount of energy were thrown into the ergosphere and then exploded into two fragments in such a way that one fragment had negative energy (and the opposite angular momentum to the hole itself), the negative-energy fragment would fall into the black hole, while the other fragment, to conserve the total amount of energy and momentum, would escape from the ergosphere with energy greater than the original energy which the complete particle had possessed at first. In principle, the energy of the escaping particle could be made to do useful work.

The effect of the infalling particle would be to *reduce* the total mass-energy of the black hole, and repeated application of the process could result in the extraction of a considerable fraction of the overall mass-energy. There is a limit. The effect of dropping in particles with opposite spin to the hole is to slow it down; when the rotation has ceased, no further energy can be extracted by this process. If we start with a black hole spinning at the maximum permitted rate (the maximal Kerr black hole), by reducing its final rotation to zero we can extract 29% of the initial mass-energy of the hole, and this is a lot of energy: nuclear processes going on inside stars are not nearly as efficient as this, less than 1% of the mass involved in these reactions being converted into energy. The Penrose process hints at the possibility that spinning black holes may be the central engines responsible for some of the most powerful energy sources in the Universe.

An important theorem regarding black holes and the possibility of extracting energy from them was proved in 1971 by Professor S. W. Hawking of the University of Cambridge, an earlier form of the result having been obtained about a year before by D. Christodolou. This, the *area theorem*, states in essence that the area of the event horizon of a black hole cannot decrease: whatever happens to the black hole, its "surface area" must either increase or stay the same as before. The surface area of a black hole depends on the square of its mass (the radius of a black hole is proportional to its mass, and the surface area of a sphere is proportional to the square of its radius). If we take the simple case of the collision and coalescence of two identical black holes, a certain amount of mass may be lost—radiated away as gravitational waves—in the process. The greatest possible mass loss would occur if the surface area of the resulting black hole were precisely equal to the total area of the original holes, and this gives us an upper limit to the energy released in such a process. Once again, the maximum energy which could in theory be released amounts to 29% of the total mass-energy of the two original black holes. In practice, such an event is highly improbable; and, if it did occur, calculations indicate that the total release of energy would be substantially less than this.

As we shall see later, black holes can be substantial energy sources. Even a nonrotating Schwarzschild hole could provide a gravitational well deep enough to be responsible for a release of energy more efficient than thermonuclear fusion. For example, clouds of matter falling in towards the event horizon of a black hole would become severely heated by collisions between rapidly accelerating particles, and would emit large amounts of electromagnetic radiation, particularly in the form of X-rays, *before* vanishing from view within the horizon. In effect, matter located at a distance from a black hole has a considerable amount of potential energy; as it falls towards the hole, the matter trades its potential energy for kinetic energy, some of which may be released as heat or in other forms of radiation, just as the kinetic energy picked up by a falling rock is dissipated when that rock hits the ground as heat, sound and shock waves.

Although black holes *are* black, and cannot emit radiation from within their event horizons, their powerful gravitational fields could make them some of the most potent energy sources in the Universe because of the energy released by infalling matter before crossing those horizons.

Inside a spinning black hole

The interior of a rotating (or of a charged) black hole is very different from the interior of a nonrotating Schwarzschild one. The central singularity takes the form of a ring, and if we plot it on a space-time map we find that—unlike the Schwarzschild singularity—it is vertical, parallel to the time direction on the Penrose diagram. In other words, the singularity is *time-like*. This implies that it is possible to take a route into the spinning black hole which avoids the central singularity and its horrific gravitational stresses. In fact, an object would have to be aimed rather carefully in order to hit the singularity of a Kerr black hole: only those which fell in along the equatorial plane would strike the singularity and be utterly destroyed.

The full Penrose diagram for a rotating black hole reveals that such a hole has two event horizons, an outer horizon and another

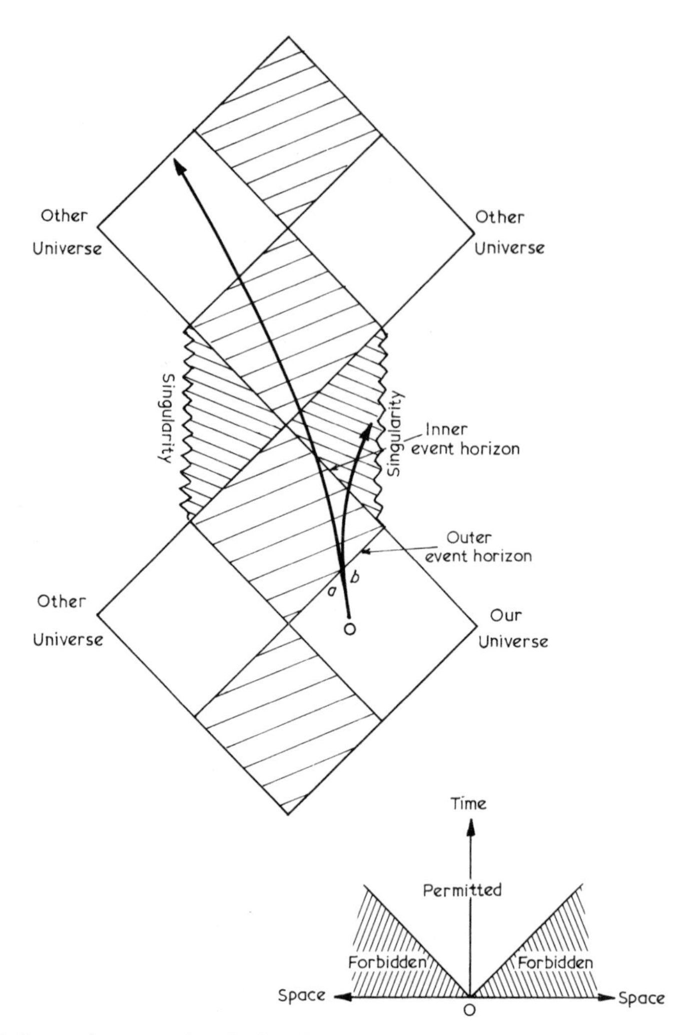

Fig. 35 *Space-time map of a spinning (Kerr) black hole.* A spinning black hole contains an outer and an inner event horizon, space-time within the inner horizon having different properties from space-time between the horizons. The singularity is time-like—i.e., it lies in the vertical direction on the map—and, as a result, a particle entering this kind of hole can *in principle* avoid hitting the singularity. The map reveals that, apparently, a spinning black hole connects our Universe with an indefinite number of "other universes"; in theory an astronaut could enter the spinning hole along a path *a* which emerged in another universe; he could, of course, choose a path *b* which encountered the singularity. There is very considerable doubt as to whether this map represents the real physical situation inside a spinning black hole. (Compare with fig. 33, page 127.)

closer to the singularity. Within the outer horizon, a body cannot prevent itself falling inwards, for on crossing that horizon, as we have already seen, the properties of space and time change so as to prevent motion taking place in any desired direction. On crossing the inner event horizon the properties of space and time change again, and it is then possible to move in directions which take the infalling body away from the singularity. We can trace the progress of an infalling astronaut in fig. 35. Within the inner event horizon he can use his rocket motors to alter his route. Without exceeding the speed of light (i.e., without taking a path inclined by more than 45° to the vertical axis) he can avoid the singularity and move away from it, eventually emerging into a new space-time. In short, after falling into a black hole in our Universe, he cannot reemerge from that hole: therefore, he must emerge "somewhere else", and that somewhere else would seem to be *another universe.*

The Penrose diagram, if worked out in full, indicates that there may be an infinite number of past and future universes. By weaving in and out of spinning black holes, an intrepid astronaut could pass from universe to universe for the rest of his life. However, if this picture is correct as drawn, he can only move to "future universes"; he cannot, by entering a black hole, come back to his own universe, and to the time of his contemporaries. Entering a spinning black hole still implies a one-way trip for the astronaut, even if it does not necessarily lead to his destruction. By the same token, astronauts from other "past universes" could enter ours by falling into spinning holes in their own universes. These bizarre and fascinating properties of the Kerr solution have aroused the widest popular interest, for the possibilities apparently opened up are mind-bending in the extreme.

To take matters further, it has been speculated on the one hand that the "other universes" are, in fact, our own Universe, and on the other that the set of universes in the Penrose diagram may be linked up in a way which would allow a route to be charted which eventually reemerged in our own Universe—*possibly in the past.* This raises the delightful possibility that by entering a conveniently situated spinning black hole an astronaut could achieve virtually instantaneous travel to different parts of our own Universe. In particular, by such means it might be possible to get from A to B by this "back door" route faster than a ray of light could travel from A to B through normal space. Indeed, it should be possible for an astronaut to make a journey and arrive home before he set out!

As we have already discussed in Chapter 4, the fundamental law of causality—that cause must precede effect—would be destroyed by such a possibility. If causality were destroyed the Universe would be a wholly irrational and unpredictable place. If faster-than-light transportation *via* the agency of black holes were possible, that would be the shattering outcome. To maintain sanity in the Universe, we have a vested interest in its not being possible to use black holes in this way.

Of course, there is one major practical objection to this mode of transportation: any spacecraft approaching close to the event horizon of a stellar-mass black hole would be torn to shreds by tidal forces. But that, of itself, does not eliminate the possibility that particles, or information, might be communicated in violation of causality. In any case, with super-massive black holes, the problem of tidal stress need not arise: astronauts venturing into, say, a billion-solar-mass hole could complete their "journey" without encountering any discomfort from this source.

The objection to travelling through spinning black holes may be much more fundamental. There is a growing body of theoretical evidence which suggests that the interior of a Kerr black hole (or of a charged black hole) is too idealized, and that in a real black hole the space-time bridge would not exist. Kerr–Newman models treat black holes as if they existed in isolation within a flat space-time, and so ignore the effects of matter in their vicinity, and also possible quantum effects, which could be highly significant. For example, in 1978 N. D. Birrell and P. C. W. Davies showed that quantum effects would destroy the space-time bridge in a charged black hole, and similar arguments are thought to apply to rotating ones, too. In practice, the singularity in a real spinning black hole may turn out to be space-like, and the fate of any infalling astronaut would be to be crushed in the singularity, just as he would be if he fell into a Schwarzschild black hole.

It is sad in a way to lose so exciting a possibility as travel through black holes. The possibility is not yet dead, for we are still far from having a full understanding of what goes on inside spinning black holes, but the current trend of research is very much against being able to use them as rapid-transit systems.

Naked singularities

In Chapter 7 we encountered the cosmic-censorship principle whereby it is proposed that naked singularities cannot exist; i.e., that all singularities in nature are surrounded by event horizons which prevents their being seen by the outside Universe. Can charged or spinning black holes have naked singularities? The equations describing the Kerr–Newman solutions show that if either the charge or the angular momentum of a hole should exceed certain critical values, then the event horizon would disappear. If we had a black hole spinning at the maximum permitted rate (or having the maximum possible charge), could we destroy the event horizon and reveal the central singularity by throwing into the hole particles which would add sufficient angular momentum or charge?

It looks as if this could not happen in practice. What matters is the *ratio* of angular momentum to mass (or charge to mass), and the process of throwing particles in through the event horizon would add just enough mass-energy to increase the mass of the black hole sufficiently to prevent the disappearance of the horizon and the emergence of a naked singularity.

If we cannot produce a naked singularity from an existing black hole, can we produce one directly from the collapse of a mass containing more than the critical amount of charge or of angular momentum? Again the answer appears to be "no". In such a situation the effect of the charge, or of the rotation, should be sufficient to prevent the formation of the black hole in the first place. For example, in the case of very rapid rotation, we could say that the "centrifugal force" generated by the collapsing matter would prevent complete collapse.

None of this is absolutely certain, but it does look as if we are unlikely to encounter naked singularities associated with gravitational collapse and black holes. Are there any other possibilities?

White holes

General Relativity is a *time-symmetric* theory. It works equally well whether time is flowing forwards or backwards; there are solutions which apply if time runs in the opposite direction to the sense in which we normally imagine it to "flow".*

*The whole notion of the "flowing" of time is a suspect one, but the fact remains that our *impression* is that time flows steadily in one direction.

There is nothing unusual about this. The same is true of Newtonian gravitation. Consider a comet moving around the Sun along an elliptical orbit. If the direction of time were to be reversed, the comet would still follow the same elliptical path: its direction would be reversed, but in every sense its behaviour would be as before—it would accelerate as it approached the Sun and decelerate as it moved away. Or, again, an ideal bouncing ball would still fall to the ground and return to your hand if time were to run backwards.

Given that General Relativity is time-symmetric, and that a black hole comes into being when a cloud of matter falls together to form a singularity surrounded by an event horizon, which then continues to exist forever (ignoring quantum effects, which will be discussed in later chapters), could there exist the time-reverse of a black hole—i.e., the collapse process with the order of events reversed? If such an entity were to exist, it would be a region of space out of which matter would suddenly begin to pour, a *white hole.* It would contain a singularity which had existed from the beginning of time but from which, at a random and unpredictable instant, matter would begin to pour. Indeed, just as particles can fall into a black hole after its formation so, presumably, could particles emerge from a white hole before the hole itself burst forth into a cloud of matter. Since particles and radiation could emerge from the central singularity of a white hole, it would constitute a naked singularity.

The possible existence of white holes has been debated for some time. Various authors have suggested that white holes may lurk at the centres of "exploding galaxies", and that other energetic phenomena such as quasars might represent regions where matter is pouring into our Universe. The concept has provided a tempting potential solution to the behaviour of some of these enigmatic objects. The Soviet astronomer I. D. Novikov suggested in 1964 that white holes might have arisen as a result of events which took place in the Big Bang when, as current theory suggests, the Universe itself emerged from an infinitely dense configuration—the initial singularity. He argued that if some regions of the initial Big Bang did not immediately participate in the overall expansion, these regions, or "lagging cores", might burst forth at some later stage as white holes.

A glance at the Penrose diagram for a spinning black hole (fig. 35, page 136) suggests another possibility, namely that if other universes

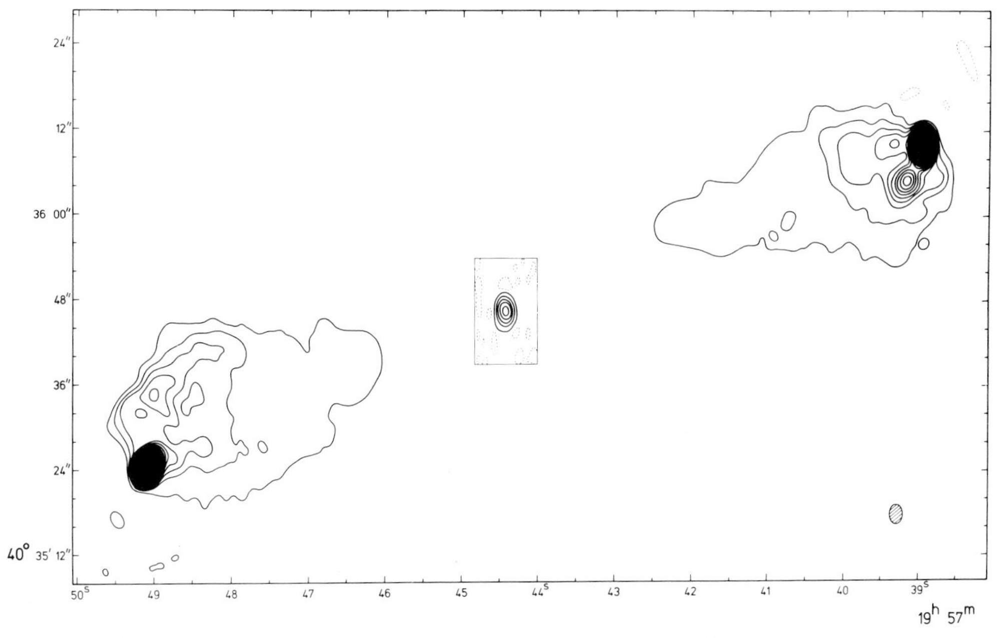

Radio map of Cygnus A showing the two principal radio-emitting regions symmetrically located on opposite sides of the central galaxy which, itself, is a relatively weak radio emitter (contained within the rectangular box). The two radio lobes, separated from each other by about 600,000 light-years, are believed to represent clouds of material ejected from the central galaxy. (Courtesy Mullard Radio Astronomy Observatory, Cambridge.)

The quasar 0420–388. An X-ray photograph taken by the Einstein Observatory, showing one of the most distant X-ray quasars; with a red-shift of 3.1, its distance may be as great as 15 billion light years. (Courtesy H. Tananbaum, High Energy Astrophysics Division; Harvard/Smithsonian Center for Astrophysics.)

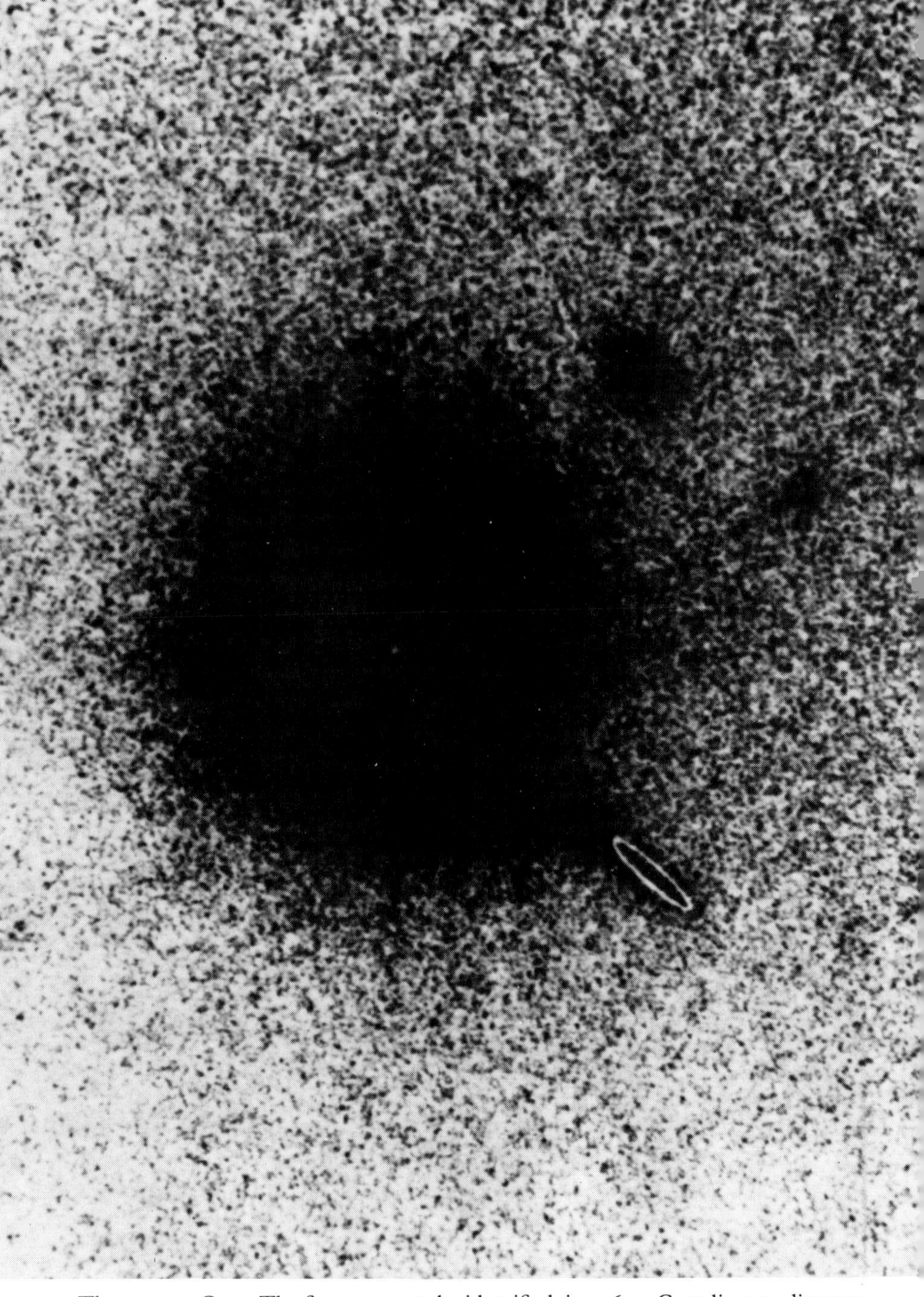

The quasar 3C273. The first quasar to be identified, in 1963, 3C273 lies at a distance of over 3 billion light years and is almost star-like in appearance. It is a double radio source, the stronger of the sources coinciding with the 'jet' and shown on this negative photograph by the white oval; the weaker source coincides with the centre of the star-like object. Although quasars are very tiny on the astronomical scale, they radiate hundreds of times as much energy as ordinary galaxies. (Photograph from the Hale Observatories.)

exist the collapse of matter into a black hole in a past universe could result in its emergence as a white hole in our Universe. In other words, "someone else's" collapsing star would become our white hole. It must be admitted that there is an appealing sense of symmetry about the idea of black and white holes, and about the possibility of there being an endless variety of universes with which we have no contact in the normal course of events except *via* the agency of black and white holes.

Indeed, black and white holes might not be the only possibilities. "Grey holes", consisting of matter which, having been spewed out of an event horizon, almost immediately recollapses again, have also been mooted.

Can white (or grey) holes really exist? We have already seen that the consensus of opinion regarding the concept of space-time bridges in spinning or charged black holes is that it is too idealized, and that such bridges would not exist in practice. Instead all black holes probably contain space-like singularities into which matter will fall. Such considerations militate against the existence of white holes arising from collapsed stars in other universes (or in different regions of space-time).

The possibilities for "lagging cores" do not seem to be too bright either. As D. M. Eardley argued in 1974, since radiation near to a black hole is strongly red-shifted, radiation in the vicinity of a white hole should be strongly blue-shifted; in the earliest instants of the Universe, when matter and radiation were densely concentrated, a tremendous build-up of strongly blue-shifted and highly energetic radiation would occur around the boundary of the potential white hole. So much mass-energy would be concentrated in a small volume that a black hole would form around the incipient white hole, cutting it off from the rest of the Universe before it had time to "pop off". As K. Lake, of the University of Toronto, and others have argued, the only way in which a lagging core could become a *bona fide* white hole is if it began its expansion with practically no delay after the Big Bang itself.

Lake suggests three possibilities: that some white holes may have recollapsed to form black holes; that some may have expanded far beyond their Schwarzschild radii, and so be unrecognizable today; and that there may be some local inhomogeneities in the Big Bang which have not expanded beyond their Schwarzschild radii and which have always been visible. Radiation can escape from a white hole, but Lake's calculations indicate that radiation from a white hole of the last kind would be very strongly red-shifted, so that the source would appear as a very dull red point rather than something dramatic and spectacular.

Hawking has argued that, if white holes exist, they will be indistinguishable from black holes; others such as Penrose have argued strongly that white holes cannot exist at all. For one thing, they offend against the cosmic-censorship principle, because the singularity is, at least in principle, visible; but is should be recognized that cosmic censorship is not an established law of nature but merely a principle which makes life easier for physicists and which seems likely to apply in a wide range of situations. Penrose has also argued, in interesting fashion, that the only reason white holes were considered at all was to preserve the notion of time symmetry (i.e., that fundamental laws of nature apply equally well if time is reversed). Since this is manifestly untrue for large-scale phenomena in the Universe (all large-scale phenomena from the evolution of life to the behaviour of the Universe itself), there is no *need* to invoke it to produce such "undesirable" entities as white holes.

Although there is still room for debate, it does appear either that white holes do not exist at all or, if they do, they may be faint and unspectacular at best, and indistinguishable from black holes at worst. It seems most unlikely that they can exist as the spectacular outpourings of matter and energy required to account for puzzling, highly luminous entities such as quasars or—as some have suggested—certain types of supernovae.

If much of this chapter seems to have consisted of setting up possibilities and then shooting them down, that is a fair reflection of the way that the climate of opinion has moved in the last few years. Of the existence of black holes themselves, however, there is very little doubt. Their existence is a natural consequence of the best current theory of gravitation—General Relativity—and of most rival theories. Even straightforward Newtonian gravitation leads to the possibility of black-hole-type objects. The theory of stellar evolution offers no tangible alternative to catastrophic gravitational collapse for a star which, at the end of its life, is too massive to become a white dwarf or neutron star. Even if some as yet unknown force prevents the final stage of collapse into an infinitesimally small singularity, this should not affect the formation of an event horizon, and the generation of that region of space which we call a black hole. As we have seen, a black hole can form with matter at surprisingly low densities, if there is enough of it available.

The case *seems* to be proven.

9
How Black is Black?

Until the early nineteen-seventies our concept of a black hole was quite clear. By definition, a simple Schwarzschild black hole was a region of space from which nothing, not even light itself, could escape. Absorbing everything which came its way and emitting nothing, in every sense it fulfilled the definition of utter blackness. Admittedly, material falling towards the event horizon of a black hole could emit substantial amounts of energy *before* finally and irrevocably vanishing from sight, but if a black hole were to exist in the depths of space, in isolation from any matter, it should emit nothing at all and be completely invisible.

The "classical" theory suggested that once a black hole had formed it continued to exist forever, or at least until the end of the Universe. A black hole could not become smaller, and it could not lose mass. Although a rotating black hole could lose its rotation over a period of time, and a charged hole could lose its charge, in each case the end result would be a nonrotating Schwarzschild black hole which would survive for the rest of eternity. A black hole could not shrink; it could only grow bigger. As time went by a black hole would absorb matter and radiation which approached sufficiently close to its event horizon: this would add to its mass and, since the radius of a black hole is proportional to its mass, the black hole would increase in size. A black hole was envisaged as a bottomless pit with an insatiable appetite for mass-energy.

The area theorem proved by S. W. Hawking (see Chapter 8, page 131) related to this aspect of black holes. The area of the event horizon could not decrease: if matter or radiation were to fall into a black hole, the area of its event horizon would increase, and if two black holes were to coalesce, the area of the resultant event horizon would be greater than or equal to the combined surface areas of the two original holes. This behaviour of event horizons is analogous to the behaviour of a quantity known as *entropy* which crops up in thermodynamics, a science which may be described, rather loosely, as "the science of heat"; more properly, it is considered to be the science of the behaviour of energy and information in physical systems—heat, after all, is a form of energy.

The famous second law of thermodynamics states that the entropy of a closed system cannot decrease; in any process which takes place, it must either increase or stay the same. By entropy we mean the "inutility" of energy—the fact that energy may not be available in a suitable form to do useful work; alternatively, we can regard the entropy of a system as being a measure of the disorder of that system, or of our lack of information of its precise state. If entropy increases, the amount of energy available to do useful work decreases, or the amount of information which we have about the internal state of a system decreases. The second law is a pessimist's charter: essentially it implies that, in the Universe as a whole, things can only get worse!

For example, if we have a bucket of hot water and a bucket of cold water, we can make use of the temperature difference between them to run a machine which will do useful work. If we pour the contents of the two buckets into one container they will mix to form water at a uniform temperature, from which we cannot extract useful work. The water still contains energy (it is warm), but we cannot extract work to run our machine any more. Or, again, let us begin with a jug of black coffee and a jug of milk, and pour the contents of both into a cup, so ending up with brownish milky coffee. In the initial state we had a certain degree of order and information: coffee was in one container and milk in the other. After the interaction we have a less ordered system: milk and coffee are mixed and we cannot locate and separate one from the other. We have lost the information which we originally possessed about the location of the milk and the location of the coffee. In both examples, the entropy of the system has increased.

In 1972 J. Beckenstein examined the similarity between the behaviour of entropy and the properties of event horizons. They share the characteristic that they never decrease, but tend to increase in any processes in which they are involved. Could the analogy be taken further to provide a meaningful link between black-hole physics (gravitation) and thermodynamics, two apparently disparate sciences? Could a black hole possess entropy and, if so, what was meant by the entropy of a black hole?

In one sense a black hole clearly has a very high entropy indeed. A black hole has only three distinguishing features: mass, charge and angular momentum. Consequently an almost unlimited number of configurations of particles could have gone into producing black holes indistinguishable from each other. A vast amount of information is lost in the formation of a black hole. Beckenstein argued that the entropy of a black hole could be described in terms of the number of possible internal states which correspond to the same external appearance. The more massive the hole, the greater the number of possible configurations which went into its formation, and the greater the loss of information. The area of the event horizon is related to the mass of the black hole (in fact, it is proportional to the square of the mass): the more massive the hole the greater its entropy, and the more massive the hole the larger the area of its event horizon, so it seemed reasonable to regard the entropy of a black hole as being proportional to the area of its event horizon.

The problem that arose with assigning a finite value of entropy to a black hole was that this implied that *a black hole should have a finite temperature.* For a black hole to have a temperature it would have to be emitting energy, an idea quite contrary to the entire basic concept of a black hole. The analogy was developed further in 1973 by J. Bardeen, B. Carter and S. W. Hawking, who showed that the surface gravity of a black hole played an analogous rôle to the concept of temperature in thermodynamics. The surface gravity at the event horizon of a black hole is inversely proportional to its mass and, if the analogy is carried through, this implies also that the temperature of a black hole is inversely proportional to its mass; i.e., the less massive the hole, the "hotter" it would be.

Around this time S. W. Hawking was investigating the quantum behaviour of matter in the neighbourhood of a black hole. To his own considerable surprise, and to the surprise of the scientific community when he published his results in 1974, he discovered that black holes would appear to emit particles such as photons, electrons and neutrinos, and that to a distant observer this radiation would have a thermal spectrum; i.e., the same kind of spectrum as one would expect to receive from an ideal hot body (an emitter of radiation known, somewhat confusingly, as a black body). Hawking's investigation of quantum effects showed that black holes would behave as if they had temperatures, and this behaviour fitted in exactly with the analogies which had been made between black-hole physics and thermodynamics. Black holes turned out to emit radiation of the same type as would any hot body, and to have temperatures which were inversely proportional to their masses. This startling discovery showed that black holes were not completely black after all, and it opened up the way for the establishment of links between gravitation—which hitherto had been out on a limb, apparently unrelated to other forces—and thermodynamics and quantum theory.

How can a black hole radiate if nothing can cross its event horizon without exceeding the speed of light and so violating one of the tenets of modern physics? The answer lies in the nature of quantum mechanics. Quantum theory implies that we cannot make precise statements about, for example, the positions and motions of particles; we can determine only the probabilities of finding particles in particular places and having particular velocities. The basis of this assertion lies in the uncertainty principle, which we met in Chapter 4 (see page 78).

According to the uncertainty principle we cannot determine the *precise* energy of a particle or a physical system over an arbitrarily short interval of time. Over a longer period we can make very accurate statements about it, but the shorter the time interval the less precise will be our knowledge. The quantity which relates the uncertainty in energy with the uncertainty in time is Planck's constant.

Combining the uncertainty principle with the equivalence between mass and energy established by Einstein ($E = mc^2$), we find that in ordinary "empty" space particle-antiparticle pairs (that is, a particle of matter and a particle of antimatter) can come into existence very briefly before coming together again and destroying themselves. Within a tiny region of space, over a microscopic interval of time, there is uncertainty as to the amount of energy which exists. Consequently a certain amount of energy can exist for a brief instant; in accordance with the uncertainty principle, the shorter the interval of time the greater the quantity of energy which can exist. If the quantity of energy is sufficiently great it can assume the form of (i.e., "create") a particle and its antiparticle, both of which will exist for a tiny instant before mutually annihilating. Provided that the annihilation takes place within the minute period of time permitted by the uncertainty relation, this process is permitted by the laws of nature. The higher the energy associated with such particles, the shorter the instant for which their existence is allowed.

Such particles are known as *virtual* particles because they cannot be observed directly; however, they do produce indirect effects which can be and have been measured. This view of empty space is now well established: far from being an empty and uninteresting void, space is a seething soup of particles fleetingly materializing and annihilating in such a way that the net matter-energy content of the Universe remains constant.

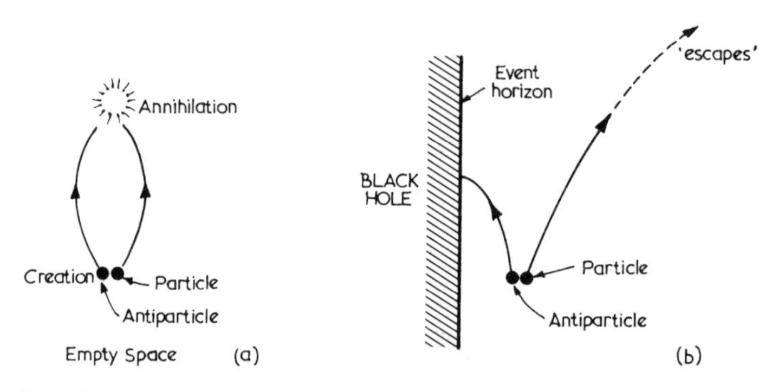

Fig. 36 *Particle creation near black holes—the Hawking process.* In so-called "empty" space (*a*) particle-antiparticle pairs are being created and, almost immediately, annihilate each other. It has been suggested that close to the event horizon of a black hole (*b*) the tidal forces may be sufficiently strong to pull apart the newly created particle-antiparticle pair so that the two cannot meet up again and mutually annihilate. In some cases the particle (or antiparticle) may escape to infinity, while the antiparticle (or particle) falls into the hole. To a distant observer it will appear as if particles are being emitted by the hole.

The powerful gravitational field close to a black hole intensifies the process of pair production. Normally we would expect these pairs to annihilate each other almost instantaneously so that there would be no net creation of particles. However, the immense tidal stresses close to the event horizon could have the effect of separating a newly created particle from its antiparticle (fig. 36). In some cases the particle and the antiparticle would both plunge into the black hole, but in other cases only one of the pair would fall in, leaving the other without a companion with which to collide and annihilate. A proportion of these could escape from the vicinity of the hole to reach distant observers, and give the impression of a flow of particles emanating from the hole. This is one way of looking at the situation, but the precise mechanism by which particles are emitted is not certain.

Whatever the exact nature of the process, its effect is to extract mass-energy from the hole just as if particles were escaping from within it—"tunnelling" out in some way. The phenomenon of "tunnelling" is a familiar one in many aspects of quantum physics. Particles faced with a barrier which, according to the laws of classical physics, they should not be able to cross nevertheless have a finite probability of getting to the other side, just as if they had "tunnelled through" that barrier. This probability can be calculated on quantum-mechanical principles.

The particles emitted by a black hole can be regarded as having tunnelled out through the event horizon by quantum-mechanical principles, so reducing the mass of the hole. The probability of a particle's escaping depends on the "thickness" of the barrier through which it has to tunnel, and this in turn depends on the mass of the hole. The more massive the hole the thicker the barrier, and the less chance there is of particles tunnelling out. The rate of emission of particles is inversely proportional to the mass of the hole, so that particles can escape more readily from a small hole than from a large one.*

As with any other "black body", the rate of energy emission (the power output) from a black hole depends on the surface area of the hole and on the fourth power of the temperature. The surface area of the event horizon is proportional to the square of the mass while the temperature of a black hole is inversely proportional to the mass; taking these two factors together we find that the power output of a black hole is inversely proportional to the square of the mass. The power output corresponds to the rate at which the black hole is losing mass; so the more massive the hole the slower the rate at which mass is being lost.

A black hole with a mass comparable to that of the Sun would have a temperature of about one ten-millionth of a degree above the

*That pairs of particles would be created by the gravitational field close a *spinning* black hole had been suggested in 1970 by the Soviet physicist Ya. B. Zel'dovich, but it had been thought the process would cease when the angular momentum of the hole was removed. Hawking showed that even a nonrotating hole would emit particles.

Absolute Zero of temperature, far too low to have a significant or detectable radiation output. There is more than enough matter and radiation floating about in the Universe at present for the mass of the black hole, as matter and energy fall in, to increase faster than the Hawking process diminishes it. Even assuming nothing whatever were to fall in, it would still require something like 10^{66} years for a solar-mass black hole to evaporate completely. Since this is a period some 10^{56} times greater than the presently estimated age of the Universe, it is clear that black-hole radiation cannot have had any remotely significant effect on stellar-mass black holes since the Big Bang. Since the time required for a black hole to evaporate completely depends on the cube of its mass, a black hole of ten solar masses would be expected to survive a thousand times longer than even a solar-mass hole!

Is the Hawking process then of purely academic interest only? Perhaps not. If black holes with very low masses were to exist, they would have much higher temperatures and would evaporate over much shorter timescales. As Hawking showed in 1971, fluctuations in density just after the Big Bang could have squeezed together relatively tiny pockets of matter to form mini black holes of small mass and microscopic size. Such holes are known as *primordial* black holes. The pressures necessary to generate black holes from small quantities of material cannot be produced in the Universe today, but in the first instants of time it is quite conceivable that this could have happened. A primordial black hole containing the mass of a small mountain (say a billion tonnes, or 10^{12}kg) would be comparable in size with a proton, and the density to which that material would have had to be compressed in order to generate a black hole would have been equivalent to the density which would be obtained if all the galaxies in the observable Universe were squeezed into a sphere about ten *centimetres* in radius!

A primordial black hole of this mass would have a temperature of about 10^{11} K. It would be emitting electrons, positrons, photons, neutrinos and other kinds of particles with a power output of some 6,000 megawatts. This power output would be equivalent to that of several large terrestrial power stations.

As a black hole loses mass, its temperature increases. The hotter it becomes the faster it radiates, and the faster it radiates the faster it loses mass. As the mass of the hole becomes very small the process escalates very rapidly until, in the end, the black hole radiates away the last of its mass-energy in a catastrophic explosion. Primordial black holes of extremely low mass would have exploded soon after their formation, but black holes of about a billion tonnes would be expected to decay over a period of about 10^{10} years, a period of the same order as the estimated age of the Universe. Some primordial black holes should be exploding now.

In our present state of knowledge it is impossible to predict precisely what would occur in the final stages of black-hole evaporation; but it is certain that the final explosion would result in the release of a tremendous burst of high-energy gamma rays. Prediction of the *amount* of energy released depends upon the theory of elementary particles adopted to make the calculations. According to one of the "simpler" current theories (if such a term can be applied to *any* theory of so-called "fundamental" particles), which suggests that all massive nuclear particles are made up of six basic particles called *quarks*, the energy released by a primordial black hole exploding now during the final second of its existence is about 10^{22} joules. This is comparable with the explosive violence of about 10 million 1-megaton hydrogen bombs. Adopting an alternative theory, due to R. Hagedorn, which predicts the existence of an unlimited variety of species of elementary particles, the resulting explosion would be fully a hundred thousand times more violent—equivalent to a thousand billion hydrogen bombs!

Clearly, if one were to use primordial black holes as an energy source, one would have to take good care to throw them well away before the time of the final explosion. An exploding primordial black hole would be, alternatively, an utterly devastating weapon of destruction. Even a mini hole with a long life still ahead of it would be a thoroughly lethal entity, pouring out thousands of megawatts of high-energy gamma rays which would do no good at all to living creatures in the vicinity.

What would be left behind after the explosion? We do not have a theory capable of explaining what happens when a black hole shrinks within the Planck radius (about 10^{-35} m), so this question lies in an area of doubt and speculation. The black hole may disappear completely, leaving nothing but the energy released in its final explosion. The evaporation might leave behind a nonradiating black hole of around the Planck mass (10^{-8} kg), or—and this is the most bizarre possibility of all—perhaps the emission of energy might continue indefinitely, leaving behind a naked singularity of *negative* mass. The consensus of opinion favours the first possibility, that the black hole evaporates completely, leaving nothing behind. Even so, in the final instant of the explosion the outside Universe would be exposed directly to the singularity at the core of the exploding hole. A borderline case for the cosmic censor!

The detection of an exploding black hole would be a discovery of the utmost importance. Not only would it demonstrate the validity of the Hawking theory, and of the links between gravitation, thermodynamics and quantum theory, but, because different theories of particle physics make wildly different predictions about the properties of such explosions, analysis of the energy emissions would provide crucial information about the nature of fundamental particles. Having said that, it is only fair to stress that no such explosions have been detected as yet, and that even if they do occur they would be very difficult to detect. There is no certainty that primordial black holes exist at all; they remain, for the moment, no more than a theoretical possibility.

Science in the late nineteenth century had made, and was making, great strides forward, and there was no idea then that there might be fundamental limits to what we can know. There was a feeling that, with diligence and patience, any quantity could be measured, and the future behaviour of the Universe and its contents could be predicted.

The advent of quantum mechanics changed that state of mind. The uncertainty principle states that we cannot simultaneously know the precise position *and* the precise velocity of a particle (the greater the precision with which we measure one, the less the precision with which we can know the other). Quantum mechanics implies that we cannot predict the precise outcome of experiments involving single particles; all that can be done is to calculate the probabilities of different outcomes. The measure of uncertainty introduced to the Universe by quantum mechanics was repugnant to Einstein, who is said to have remarked: "God does not play dice."

The emission of particles from black holes introduces a greater degree of uncertainty, for we cannot predict *either* the position *or* the velocity of an emitted particle. Hawking argues that, in a sense, the radiation can be considered to have come from the singularity, a region where space-time as we know it breaks down. Accordingly, new random information is entering the Universe from regions of space-time of which we have no knowledge, and this reduces still further our ability to make predictions about the future state of the Universe. There is, then, what has been termed the *Principle of Ignorance*. To Einstein's remark Hawking has replied, "God not only plays dice, but also sometimes throws them where they cannot be seen."

We live in a world of probabilities, and if Hawking's ideas on black holes and quantum mechanics are correct there are fundamental limits to what we will ever be able to know or to predict. Black holes will have altered our whole philosophy of nature.

10
Black Holes in the Universe

Black holes may turn up in a wide variety of astrophysical contexts. We may expect to find black holes ranging from 2 or 3 solar masses up to about 100 solar masses formed from collapsed stars; black holes of up to a few thousand solar masses may exist in the cores of massive globular star clusters; while it has been postulated that supermassive black holes ranging from a few million to a few billion solar masses may reside in the nuclei of galaxies, particularly hyperactive ones such as radio galaxies and the enigmatic quasars. At the other end of the scale there is the possibility that primordial black holes of masses ranging upwards from a billion tonnes may be scattered around in unknown numbers. It has even been suggested that vast amounts of invisible material may be locked up in black holes roaming through intergalactic space. If large numbers of black holes exist, particularly between the galaxies, their net gravitational influence could exert a profound effect on the evolution of the Universe, a possibility which is explored further in the next chapter.

Black holes have been invoked to "explain" such a wide range of astronomical phenomena that there is a danger of their being regarded as a solution to all the unsolved problems in astrophysics. Convincing cases for the involvement of black holes can be made in a variety of circumstances, but it should be borne in mind that at the present time there is no *conclusive* proof that black holes exist at all.

Searching for black holes

We should not expect it to be easy to find black holes. Isolated black holes of stellar mass or greater would be very black indeed. If a starship should happen to be travelling directly towards one (a most improbable situation) it could easily plunge in without being aware of the hole's presence until it was too late. Shining a radar or laser beam ahead would not help, for obvious reasons.

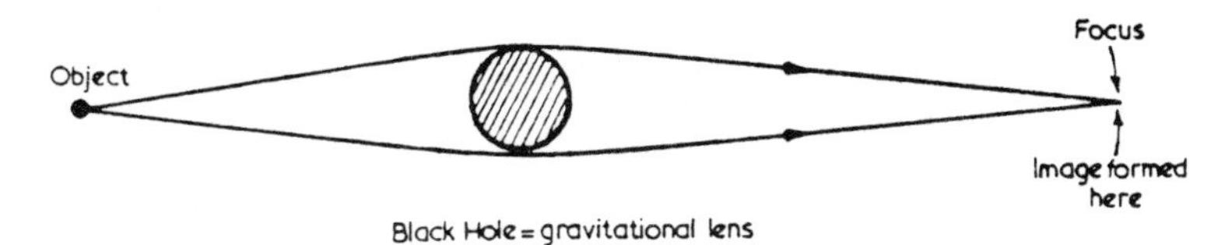

Fig. 37 *The gravitational lens.* Rays of light passing close to a massive body (e.g., a black hole; but any massive body will act as a lens of sorts) will be deflected such that, at a particular point on the opposite side of the body from the source of light, they will be focussed and an image produced. (Compare with fig. 24, page 95.)

One possible approach would be to look for the *gravitational lens* effect (fig. 37). As we saw in Chapter 5, light is deflected by passing close to a massive body in much the same way as it is bent in passing through a lens. The amount by which a light ray is deflected depends upon the mass of the body and the distance at which the ray passes. Rays of light could be brought to a focus by a black hole, and an observer located at the focus would see a magnified image of a distant object lying directly behind the hole. In the ideal situation of a point source of light and a point-mass gravitational lens (a black hole is small enough to be considered a point), the image also would be a point, like a star seen in a telescope; if the source were an extended one it would appear as a ring or, if the alignment between source, lens and observer were not perfect—i.e., if the "lens" were off-axis—as two crescents.

If hypermassive black holes—in excess of about a thousand billion (10^{12}) solar masses—were to exist in the Universe in significant numbers, gravitational lens effects would be detectable in the distribution of galaxies in the sky, particularly among the very distant quasars and radio galaxies. The lack of marked effects of this nature suggests that black holes on this scale are very uncommon, if they exist at all. There is now good evidence for one case of image-doubling involving a quasar, but it is thought that the lens in this case is a massive galaxy lying between us and the quasar: any massive body can act as a lens, but the more symmetrical it is the better the "optical quality" of the lens.

It has even been suggested that future astronomers could use the Sun as a gravitational lens to study distant objects. In 1979 Dr Von R. Eshleman of Stanford University proposed that an observatory stationed at about fifty times the distance of Pluto would lie at the focal point of rays of light from distant sources passing the edge of the Sun. In principle, enormous magnifications could be achieved. Another feature of the gravitational lens would be its ability to collect and concentrate light, so rendering visible, faint sources which, otherwise, would be undetectable. Perhaps in the not too distant future, gravitational-lens astronomy will provide yet another new "window" on the Universe about us.

However, for the moment we have not been able to use the gravitational-lens effect to locate any black holes. Are there any other approaches to the problem? The crucial characteristic properties of black holes which could assist us are as follows: they are massive (primordial mini holes excepted); they are compact; they emit nothing detectable from within their event horizons (mini holes excepted again), and so cannot be seen directly; and, because of their powerful gravitational fields, they may be powerful sources of energy. In light of these, our best hope of finding a black hole is to look for the influence which it exerts on its surroundings, and this will most readily be apparent when it is in the vicinity of matter. We should be looking for their effects on, say, the orbits of neighbouring stars, and/or for highly energetic, very compact energy sources.

Stellar-mass black holes

How may we attempt to detect black holes of mass up to about 100 solar masses?

More than half of the stars in our Galaxy are contained in binary or multiple systems, and we have already seen that some binaries contain white dwarfs or neutron stars: it seems reasonable to suppose that we may find binaries containing black holes. If one member of a binary pair should evolve more rapidly than the other and end up as a black hole, we would be left with a normal star and an invisible black hole in orbit around each other. There are examples which appear to be orbiting under the influence of an invisible companion, but it usually turns out to be a white dwarf or a neutron star too faint to be seen against the brilliance of the normal star.

AP Librae. A BL Lacertae object which, although looking like a star, is known to be an intense, compact and variable source located at the centre of a distant galaxy. A close inspection of the object, located centrally, near the top of the photograph shows it to be fuzzy round the edges due to the light emitted by stars in the inner parts of the surrounding galaxy. Like quasars, BL Lacertae objects are believed to be highly energetic compact objects located at the centres of galaxies, and there is a strong body of opinion which supports the view that they are powered by super-massive central black holes. (Photograph taken by the UK Schmidt Telescope Unit. Courtesy Photolabs, Royal Observatory, Edinburgh.)

The Seyfert galaxy, NGC 4151, showing the brilliant nucleus and relatively faint spiral arms typical of galaxies of this type. Astronomers are anxious to discover why Seyferts have such compact, active and variable nuclei. (Photograph from the Palomar Observatory, California Institute of Technology.)

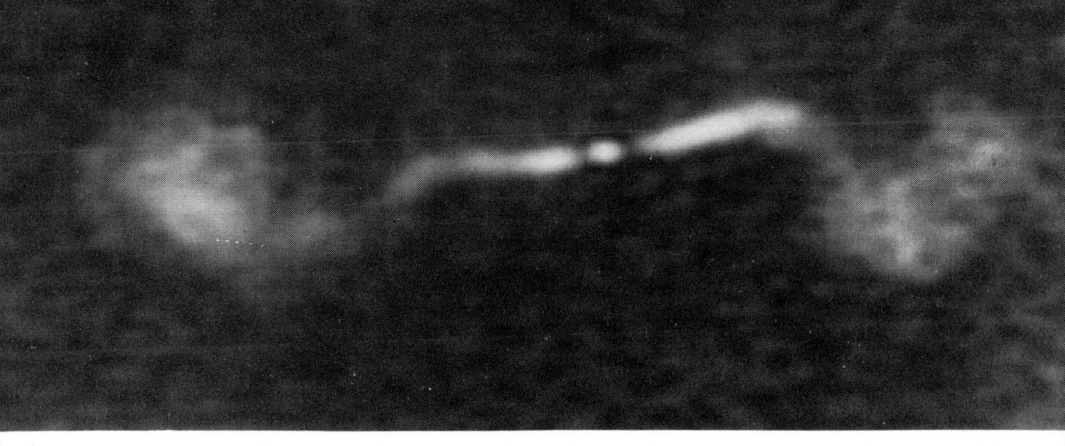

Radio photograph of the radio galaxy 3C 449 which shows this source as it would appear if the human eye were sensitive to radio waves. The central source coincides with an elliptical galaxy from which, presumably, the radio-emitting lobes were ejected, the larger and more massive lobe lying closer to the galaxy. Many double-lobed radio galaxies have a central source which, at much higher resolution, turns out also to have a double-lobed appearance, on a much smaller scale. In such cases it would appear that further ejection of material has taken place much more recently than the event which gave rise to the large lobes. It is often the case that the small-scale lobes lie along the same axis as the large-scale lobes. (Courtesy Mullard Radio Astronomy Observatory, Cambridge.)

The giant elliptical galaxy M87 in the constellation of Virgo. This short-exposure photograph shows the 'jet' emanating from the central core of this galaxy; the galaxy is a powerful radio source, and is otherwise known as Virgo A. One of the most massive galaxies known, M87 may – according to certain recent observational results – contain a super-massive black hole of some 5 billion solar masses. (Lick Observatory Photograph.)

Most binaries are spectroscopic (see Chapter 6), the stars involved being too close together to be seen as individuals; the presence of an invisible companion can be inferred in such cases from periodic changes in the wavelengths of the visible star's spectral lines, which are indicative of the periodic motion of the star around the centre of mass of the binary system.

As we saw in Chapter 6, this motion can be analysed to give reasonable estimates of the masses of the bodies involved. In a single-line binary (where the companion star is too faint to be detected) it is usually possible to make an estimate of the mass of the companion based on a knowledge of the properties of the visible star. The invisible companion usually turns out to be a low-mass star (which has very low luminosity), a white dwarf or a neutron star, but if we should happen to come across a single-line binary with an invisible companion with a mass appreciably greater than the maximum permitted mass for a neutron star, then we may begin to suspect that the invisible companion is a black hole. An ordinary star of, say, 10 solar masses would be highly luminous, and it is difficult to account for so massive an invisible companion other than by assuming it to be a black hole. However, evidence of this kind alone would be quite insufficient to ensure a "conviction". The concept of a black hole is so important, fundamental and exciting that evidence of the highest quality would be essential to constitute acceptable proof. Corroborative evidence would be essential.

If the star and its black-hole companion were sufficiently close together, the star would be distorted by the gravitational attraction, and significant flow of matter from the star towards the hole would take place. As we saw in Chapter 6, if a star is large enough to overflow its Roche lobe (fig. 28, page 107) substantial quantities of matter will flow through the inner Lagrangian point to enter orbit around the companion. Even if the star has not quite filled the lobe, a strong stellar wind will carry material away from the star and some of this will be captured by the companion.

If that companion is a black hole, what happens to the matter captured from the star? The orbital motion of the two bodies around each other will ensure that the matter does not fall straight into the hole. Instead it will go into orbit around the hole, forming a flattened disc of gas known as the *accretion disc*, the gas revolving at higher and higher speeds the closer it gets to the event horizon. At the outer edge of the accretion disc the gas will have a temperature similar to that of the star's surface but, further in, frictional heating will push up temperatures to enormous values; in the inner parts of the disc, we can expect them to exceed one hundred million degrees. The gravitational field of a black hole is so powerful that matter falling straight in is travelling at a large fraction of the speed of light by the time it approaches the event horizon. Since a black hole is very tiny, it is hardly surprising that material funnelling in towards the hole at these speeds becomes very hot indeed.

Viscosity in the accretion disc causes material to drift in towards the hole, for when particles collide some lose angular momentum and fall closer in. Material orbiting close to the hole at a large fraction of the speed of light finally spirals in through the event horizon and is lost from view (within a distance of about 3 Schwarzschild radii there are no stable orbits around a Schwarzschild black hole). It is expected that the infalling material in the disc will release copious amounts of energy amounting to, perhaps, 10% of the energy equivalent of the matter involved; i.e., about 10% of mc^2.

There is a limit to the amount of matter which can fall into a black hole in this way because, if the luminosity generated exceeds a critical value known as the Eddington limit, the pressure due to this radiation will be sufficient to blow away most of the infalling material.

Black-hole candidates

Cygnus X-1 is a powerful source of X-rays discovered in 1970 by the NASA satellite *Uhuru* which made the first full X-ray survey of the sky. The following year it was shown that Cygnus X-1 coincides in position with a hot blue supergiant star known by its catalogue number HDE 226868. Investigations carried out notably by C. T. Bolton of the David Dunlap Observatory showed that HDE 226868 is a single-line binary with an orbital period of 5.6 days. The visible star has a mass of between 20 and 30 solar masses and a temperature of about 25,000 K, making it a highly luminous star, readily visible at a range of some 8,000 light-years. Assessments of the mass of the invisible companion have ranged between 5 and 15 solar masses, but detailed investigations of the way in which the light output of HDE 226868 varies—these variations, due to the distortion of the star, were first studied in the early nineteen-seventies by V. M. Lyutiy—have allowed its probable mass to be pinned down to between 8 and 11 solar masses. Such a value is well in excess of the maximum permitted mass for a white dwarf or a neutron star, and a good case can be made for the presence of a black hole.

The presence of the X-ray emission fits in well with such a picture, since it strongly suggests that Cygnus X-1 contains a compact accretion disc from which the observed X-rays are coming. An important point about this emission is its rapid variability; the intensity of the observed X-radiation fluctuates on timescales ranging from milliseconds to months, or even years. Since no information can be communicated faster than the speed of light, a source of radiation cannot change in overall brightness by a significant amount in a time less than the time required for a ray of light to cross that source. A fluctuation in a period of 1 millisecond would thus imply that the source responsible cannot be bigger than some 300km across.

The precise mechanism responsible for the emission of radiation from the accretion disc and for the fluctuations is uncertain—quite possibly there are several different mechanisms at work. Among the possibilities under discussion are hot spots due to instabilities in the disc. For example, a hot spot orbiting in 1 millisecond would be just above the event horizon; many orbiting hot spots would give rise to chaotic periodicities, and apparently random observed fluctuations. Whatever the mechanism, the important point about these variations is the very small size implied for the emitting region.

Of course, X-ray-emitting accretion discs are expected where neutron stars exist in close binary systems, too, and many of the X-ray "stars" so far detected can be accounted for in this way. However, it does seem as if the X-ray spectrum and the intensity variations of Cygnus X-1 differ quite markedly from the behaviour associated with accreting neutron stars.

The case for Cygnus X-1 containing a black hole rests on the following evidence: the presence of an invisible companion apparently too massive to be a white dwarf or a neutron star; and the presence of an X-ray-emitting accretion disc, and the rapid variability of the emission from this. Alternative theories have been proposed—for example, it has been suggested that it is a triple-star system containing an accreting neutron star orbiting one of the other two—but such alternatives seem to be rather contrived, and there is no convincing evidence for the extra periodic effect which would be expected if three stars were involved. The simplest solution which best fits the available evidence is that Cygnus X-1 truly contains a black hole, and, ten years after its discovery, it remains the best black-hole candidate.

Did the ancient Chinese see the event which led to its formation? This tantalizing possibility was raised in 1979 by Li Qui-bin of the Peking Observatory when he pointed out that Chinese records seemed to indicate that there was a "guest star" in October 1408 in a position which corresponds quite closely to the actual location of Cygnus X-1. Was this "guest star" a supernova associated with the formation of the Cygnus X-1 black hole? The position of the reported object was such that it would have been nearly overhead in the evening skies of Europe, yet there are no European records of this event. Surely it could not have been missed, if it actually happened? There is considerable doubt, therefore, about the validity of the Chinese records, but it should be borne in mind that the Europeans did not seem to record the Crab supernova of 1054 either.

The final collapse of a star to form a black hole may or may not be accompanied by a supernova-type explosion. Theory is unclear on this point. Recent work by the Soviet astrophysicist Iosef Shklovskii suggests that one well known supernova remnant may have a central black hole rather than the central neutron star associated with objects like the Crab nebula and the Vela remnant. The object concerned is a powerful radio source known as Cassiopeia A, one of the first radio sources to be discovered and the brightest in the sky at long wavelengths. It has proved to be an expanding shell of gas, the high rate of expansion implying that a supernova explosion was involved. From the presently estimated size of the shell it can be shown that the explosion must have occurred about 300 years ago in the latter part of the seventeenth century. Even at its distance of some 9,000 light-years a normal supernova should have appeared as bright as the brightest stars in the sky and, bearing in mind that this was a period of intense astronomical activity, it would have been surprising in the extreme if the supernova had gone unnoticed. In fact, supernovae were recorded in 1572 (in Cassiopeia) and 1604; if there had been one visible in the latter part of the seventeenth century, it is almost certain to have been observed, particularly as it would have occurred in Cassiopeia, a well known constellation always visible in northern Europe.

Shklovskii suggests that the absence of a central neutron star might imply that the object which gave rise to Cassiopeia A was a very massive star whose core collapsed directly to form a black hole, and argues that any supernova event associated with this would not have been bright enough to be seen. Observations made by the X-ray satellite HEAO-2 (the "Einstein Observatory", as it has come to be known) show no indications of a central neutron star, and a neutron star only 300 years old would be expected to be a readily detectable X-ray source. A further piece of suggestive evidence is that the spectrum of the expanding cloud does not reveal traces of heavy elements like iron which we would expect to find in the remnants of a supernova (since in such explosions much of the deep interior, where heavy elements are generated, are hurled into space): this absence would seem to rule out the possibility of complete disruption of the star. If the star was not completely destroyed, yet no neutron star was formed, the remaining logical possibility is that the major part of the star collapsed into a black hole. It is too early to pass judgement on the validity of the suggestion, but it is, perhaps, worthy of note that Einstein Observatory results appear to show a surprising scarcity of hot X-ray-emitting neutron stars in supernova remnants, a fact which may necessitate some revision of our ideas about how neutron stars form and cool down.

Although Cygnus X-1 remains the best black-hole candidate, there are other sources for which a case can be made. Probably the next best candidate is another X-ray source, Circinus X-1 (otherwise known as 3U 1516-56, its number in the Third Uhuru Catalogue of X-ray sources). It, too, is associated with a binary system, and the X-ray emission displays rapid variability. Like Cygnus X-1, Circinus X-1 has been studied at X-ray, optical, infrared and radio wavelengths. Further away than Cygnus X-1, lying at a distance of some 25,000 light-years, the visible star is heavily dimmed by the effects of interstellar dust and to date it has not been possible to establish the masses of the star and the invisible companion, which revolve around each other in 16.6 days. The similarity in properties is striking, but the evidence is not so compelling as for Cygnus X-1. It is interesting to note that Circinus X-1 lies close to a possible extended supernova remnant perhaps some 100,000 years old.

Among the various other possibilities is GX399-4 (4U 1658-48) which was proposed as a candidate in 1979 by a group based at the Naval Research Laboratory, Washington D.C., on the basis of its X-ray variability. The optical counterpart has been identified, but at the time of writing insufficient data are available to show whether or not a binary system was involved.

Since the latter part of 1978, the astronomical community has been fascinated, perplexed and excited by a truly weird object labelled SS 433 (number 433 in a catalogue of stars produced by C. Stephenson and N. Sanduleak of Case Western Reserve University, Ohio, the distinguishing feature of these stars being strong emission lines in their spectra). SS 433 is a strong radio source, showing similar radio variability to Circinus X-1, and lies close to the centre of a supernova remnant known as W50. Using the Anglo-Australian Telescope P. Murdin and D. Clark of the Royal Greenwich Observatory had independently identified a point radio source that appeared in the Molongolo–Parkes radio survey with a star displaying emission lines of hydrogen. They proposed that these lines were emitted by gas flowing onto a collapsed object in a binary system at the centre of the supernova remnant. This source, which turned out to be SS 433, was also detected as an X-ray source by the British satellite Ariel 5, launched in 1974.

Then came the most startling discovery of all. B. Margon of the University of California at Los Angeles found that against a background of stationary emission lines there were pairs of lines which moved to and fro in wavelength with a period of 164 days. To explain the amount by which the lines shifted in wavelength by the Doppler effect required the clouds of matter responsible for this emission to be moving at colossal speeds: the radial velocities (i.e., motions towards and away from us) alone turned out to be between 30,000 and 50,000km per second, between one tenth and one sixth of the speed of light.

What could this object be? M. Milgrom and his group in Israel, and R. J. Terlevich and J. E. Pringle in Cambridge (England), were among those who proposed massive black-hole models, the radiation responsible for the moving emission-line features being assumed to come from material circulating a hole of between a few hundred thousand and a few million solar masses. If the material were circulating in a ring, one side of the ring would be seen to be approaching while the other was receding.

Arising from work by Margon and G. Abell in the USA, P. G. Martin and M. Rees in the UK, and others, an alternative picture is now emerging. The moving emission lines probably originate in two oppositely directed jets of material moving out from a spinning central source at speeds of about 80,000km per second (about 27% of the speed of light); as the axis of the source precesses, like a wobbling spinning top, the jets periodically point more or less directly towards and away from the Earth (in fact, they always point at an angle to the line of sight: see fig. 38). It is the motion of these jets which gives rise to the changing wavelengths of the emission lines. There is a parallel here with the behaviour of a spectroscopic binary, but, of course, the velocities involved are enormously greater.

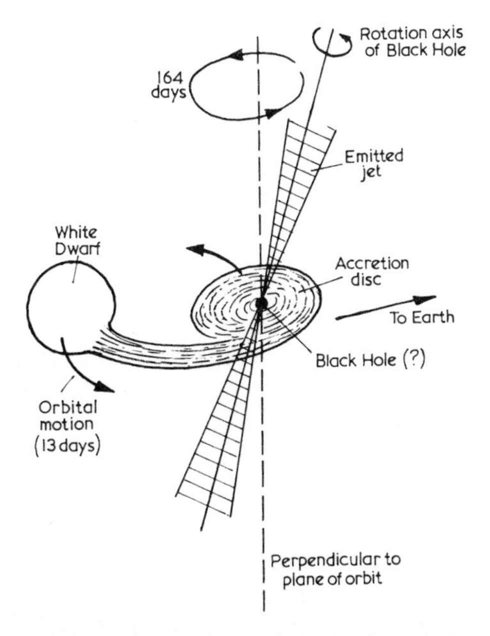

Fig. 38 *Possible model of the peculiar source SS 433.* The nature of this object is discussed in the text. One possibility, proposed by P. G. Martin and M. J. Rees of Cambridge, is that it consists of a black hole and a white dwarf in a binary system. Matter from the white dwarf is being drawn into a circulating disc of hot gas—an accretion disc—around the black hole; from the central regions of this disc matter is being expelled in two narrow jets. The white dwarf and the suspected black hole revolve around each other in a period of 13 days, and the jets precess, or "wobble", around the perpendicular to the orbital plane with a period of 164 days.

Where do the jets of material come from? Spectroscopic observations show that the "fixed" emission lines display a small periodic shift in wavelength suggesting that SS 433 is a binary, with a 13-day orbital period, containing a luminous star and a collapsed object which may be a black hole of at least 5 or 6 solar masses. The star, which is probably a white dwarf, provides the visible light and the "fixed" emission line features; gas torn from the star into an accretion disc provides the X-rays; and the jets of material are assumed to be ejected along the rotation axis of the supposed black hole. A mechanism whereby an underlying black hole could be responsible for the ejection of twin beams of high-velocity material from a hot circulating cloud of plasma was discussed by R. D. Blandford and M. Rees in 1974 in the context of radio galaxies and quasars, and we shall have more to say about that model later. However, the point is that it is quite reasonable to expect the presence of two jets of material emanating from the vicinity of an accreting black hole.

There is, then, considerable circumstantial evidence for the proposition that SS 433 may contain a black hole, but other possibilities have been suggested. For example, D. S. de Yound and G. Burbidge have proposed a model involving gas clouds trapped in strong magnetic fields above the poles of a white dwarf, and they suggest that this could explain the observed appearance in a more prosaic way. Again, in early 1980 D. Crampton, A. P. Cowley and J. B. Hutchings of the Dominion Astrophysical Observatory (Canada) published spectroscopic results suggestive of the components of the binary each being of mass less than 2 solar masses, in which case an accreting neutron star may be involved. The whole question is still very much an open one.

Clark and Murdin regard SS 433 as being a black-hole system formed in the supernova explosion responsible for the radio-emitting remnant W50, and point to similarities with Circinus X-1. If correct, this proposal would establish interesting links between supernovae and black holes and would show that neutron stars are not the only possible remnants of these events. The search is now under way for similar types of objects.

Does the Sun have a black hole companion?

On the basis of pulsar data, E. R. Harrison suggested in 1977 that the Sun may have a companion—i.e., that it may be a component of a widely separated binary or it may have made a temporary association with another object of stellar mass. Data on the periods of selected pulsars appeared to show a net distribution of frequencies which could be interpreted in terms of the Doppler effect if the Solar System were subject to a slight acceleration due to the gravitational attraction of a hitherto undetected massive body. In 1978 the suggestion was taken further by S. Pinealt of the University of British Columbia who hinted that, if such a companion were to exist, it would have to be a neutron star or a black hole, since a low-mass star or dim white dwarf of the appropriate mass and distance should have been detected before now in infrared sky surveys. The amount of acceleration, if it exists at all, is very small—about 10^{-8} metres per second per second (one billionth of the acceleration due to gravity at the Earth's surface)—and its direction would imply that the supposed companion lies somewhere within a large area of uncertainty centred on the direction of the constellations Aquila and Ophiuchus.

Pinealt's analysis suggests that Harrison's figures are consistent with a range of models, ranging from a 1-solar-mass neutron star at a distance of about 800 astronomical units to a 150-solar-mass black hole at a distance of some 9,000 astronomical units (about 50 light-days). If there is a companion, its association with the Sun is likely to be temporary, the result of a chance close encounter. An intriguing proposal made by Pineault is that gravitational lens effects should be detectable as this hypothetical collapsed body passes in front of background stars, and that a significant number of events of this kind would take place each year.

It is a fascinating possibility, surely worth investigating. The discovery of a black hole so close to the Solar System would open up the possibility of its being practicable with the technology of the 21st century to despatch a probe which might reach that hole within a relatively short period.

However, we should not place our hopes on this highly remote possibility, for there are too many uncertainties in the data and its interpretation. For the moment our best hope of detecting stellar-mass black holes remains with X-ray-emitting binaries containing massive invisible companions, such as Cygnus X-1, Circinus X-1 and, perhaps, SS 433. We have a tantalizing body of circumstantial evidence, but certainly not enough to prove conclusive.

Of maxi black holes, galactic nuclei and quasars

In principle there could exist black holes containing millions or even billions of solar masses. If they do exist, how can we expect to find them, and where should we be looking? As with stellar-mass holes, we should be looking for the effects of their powerful gravitational fields, which would show up, for example, in the motion of stars in their vicinity. Indeed, with supermassive black holes, entire galaxies could be tidally distorted. We should expect supermassive holes accreting material from their surroundings to be potent energy sources which, on the astronomical scale, would be very compact indeed, for the Schwarzschild radius of a black hole containing 50 million solar masses is equal to the radius of the Earth's orbit, while a black hole containing an entire galaxy like our own would have a radius of only one thirtieth of a light-year. Gravitational lens effects are another observational aspect worth investigating.

The best course for black-hole hunters would seem to be to look for powerful compact energy sources resulting from accretion onto massive holes. Matter is more densely concentrated inside galaxies than in the space between them, and within galaxies the greatest concentration of matter is to be found in the central regions. The ardent black-hole hunter may well regard galactic nuclei as the most fruitful hunting ground. Is there any evidence to support this assertion?

During the past two decades an astonishing variety of puzzling sources of light, radio waves and other radiations have been shown to lie far beyond the boundaries of our own Galaxy. These objects share a common need for a compact energy source capable of supplying far more power than the entire output of conventional galaxies such as our own. Among the species in the astronomical zoo with this requirement are strong radio galaxies, quasars, BL Lacertae objects, Seyfert galaxies, N-type galaxies, exploding galaxies and other peculiar objects, all generally considered to be galaxies containing hyperactive central nuclei. The light from a normal galaxy comes mainly from billions of stars spread out over a region of space something like 100,000 light-years across; the radiation which causes the striking appearance of the sources catalogued above does not come from thermal sources like stars—some other mechanism is required.

Radio galaxies are strong sources of radio waves; they were first identified with optical galaxies in 1948. In that year W. Baade and R. Minkowski of the Palomar Observatory showed that Cygnus A, the brightest radio source in the sky, coincides in position with a peculiar galaxy whose "double" appearance suggested to these investigators that they were seeing a pair of colliding galaxies. Cygnus A lies at a distance of about one billion light years, and its radio output is prodigious—comparable to the energy output of a thousand billion suns.

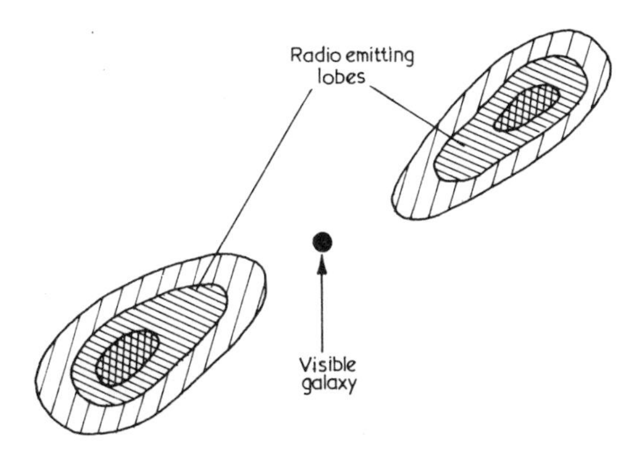

Fig. 39 *The structure of a typical radio galaxy.* The central galaxy visible in optical telescopes often has two giant radio-emitting regions, one on either side, covering a much greater region of space than the visible galaxy itself. The radio map shows contours of equal radio intensity, the greatest intensity of signal being concentrated within the innermost contours, as in this idealized example.

Since that time many more radio galaxies have been discovered, some having a source of radio emission concentrated in the centre of the galaxy, but most exhibiting a characteristic "double lobe" appearance, with two radio-emitting regions on opposite sides of the visible galaxy at distances of up to ten million light-years (fig. 39). A number of these double sources were found to have a compact central source, pouring out prodigious amounts of energy from a region less than a few tens of light years across: indeed, measurements by the technique called very long baseline interferometry have shown that in some cases the central radio source is considerably less than one light-year across.

The radio emission from the lobes is of a type known as synchrotron radiation, which is characteristic of highly energetic electrons moving at large fractions of the speed of light under the influence of magnetic fields. A great deal of energy has to be supplied to these electrons, and it is widely believed that the observed radio lobes consist of clouds of material ejected from the nuclei of the central galaxies. Clearly some kind of compact energy source is involved here.

Quasars—the term is short for *quasistellar radio sources*—are probably the most energetic and the most puzzling of all astronomical phenomena. Although we have been aware of their existence for about twenty years, there is still considerable dispute about their true nature. Despite the fact that a plausible model to explain their behaviour is beginning to gain widespread acceptance, there are still many alternative and opposing theories, and no one would claim that the mystery has been solved to everyone's satisfaction.

The first quasar to be identified was an object known as 3C 273 (its number in the Third Cambridge Catalogue of radio sources). In 1962, C. Hazard and his colleagues, using the 64m radiotelescope of the Parkes Observatory in Australia, were able to pinpoint the position of the source with great accuracy by timing the instant at which the edge of the Moon passed in front of it—and in fact they found that the source is double. The source was identified with a star-like point on a photographic plate taken with the 200in (5m) telescope at Mount Palomar; it was shown that the weaker of the two radio components coincided with the star-like object, while the more intense lay towards the end of a thin jet of light protruding from the "star".

The astronomical bombshell which brought quasars to the forefront of astronomical research was dropped the following year when M. Schmidt of the Hale Observatories studied the optical spectrum of 3C 273 and found a red-shift of 0.158 which, if interpreted as a Doppler shift, implied that the quasar must be receding from us at 15% of the speed of light. If it were assumed that 3C 273 lay beyond the fringes of our Galaxy and that its speed of recession was due to its sharing in the general expansion of the Universe as revealed by the galaxies, then it must lie at a distance of nearly 3 billion light-years. If that were so, then in order to appear as bright as it does, it would have to be emitting power at a rate of 10^{40} watts, an energy output hundreds of times greater than the output of a galaxy like our own. Yet, despite this phenomenal brilliance, the quasar was very small. Direct measurements have shown that the source 3C 273 B (the one coinciding with the star-like body) is itself double, on a scale of just a few light-years, and brightness variations have revealed that the main energy source is probably less than one light-year across. How can so small a source give out as much light as a hundred galaxies?

Many more quasars have been discovered since then and, although not all of them are radio emitters, they share the characteristics of being compact, having large red-shifts, and, in many cases, showing significant brightness variations over timescales of a year or less. Most quasars emit strongly in the infrared part of the spectrum, and recent observations by the Einstein Observatory have shown that at least 60 quasars are powerful X-ray sources too. One of these X-ray quasars, OX 169, varies in X-ray intensity by a factor of 2 or 3 within a period of some 100 minutes, which implies that the main energy source must be less than half the size of the Solar System. Clearly a very special kind of compact energy source is required for quasars.

Not all astronomers were happy at first (and some are still unhappy) with the idea of quasars being at such great distances. One school of thought suggested that they were local objects, possibly expelled from our Galaxy and rushing away at high speeds. If this were so, then they would not need to be such powerful emitters of radiation. On the other hand, it would seem odd that our galaxy alone should be the source of these objects: if the quasars were expelled from the centre of the Galaxy, they would have to be at considerable distances now in order to appear to be fairly uniformly distributed round the sky. If "our" quasars are at such distances from the Galaxy, then surely we should expect there to be quasars ejected from other galaxies, too, and to find some of them showing a blue-shift as they rushed towards us. There is something uneasily geocentric about a theory which suggests that we are at the unique centre of a major system of astronomical sources.

Gravitational—as opposed to recessional—red-shifts have been invoked as another possible explanation, but this mechanism has been shown to be unsatisfactory in explaining *all* of the observed red-shift. Even if it explains part of the red-shift, we are still left with a large residue of red-shift due to cosmological recession, and the prodigious energy problem remains.

Surveys of the distribution of quasars support the idea that they are truly at very large distances, and the great majority of astronomers accept that this is so. There are some anomalous results, nonetheless. For example, H. Arp and others have found examples of quasars so close in the sky to ordinary galaxies as to render it highly improbable that this is just a line-of-sight effect; however, the quasars have red-shifts very much greater than those of the galaxies. Perhaps these quasars are indeed associated with the galaxies, and the *difference* in red-shift is due to gravitation or some hitherto-unknown process.

The red-shifts, taken purely at face value, would indicate that the nearest quasar is about 800 million light-years away and the most distant so far discovered about 16 billion light years away. However, when dealing with such large values as the latter, our interpretation of distance is affected by the overall geometry of the Universe, as well as potentially large errors of observation, and we should not take them too literally.

There is another and most puzzling aspect of some quasars. The central radio components (i.e., those contained within the central region of diameter perhaps a few tens of light years) of some, including 3C 273, appear to be expanding, or separating, with apparent velocities greater than the speed of light. In the case of 3C 273 an expansion velocity of about $3c$ has been inferred, and in the case of 3C 84 velocities as high as $8c$ have been deduced. If these velocities are real then we are witnessing the violation of one of the most fundamental tenets of Relativity theory—and this would undermine the whole of modern physics.

Is there any other explanation? Numerous possibilities have been suggested, ranging from assuming quasars to be nearer than their red-shifts suggest, so that the measured *angular* rate of separation corresponds to a smaller *linear* rate of motion, to geometrical arrangements whereby sources moving apart with velocities close to, but less than, the speed of light can *appear* to be separating at faster-than-light velocities.

It is easy to envisage a situation in which apparently faster-than-light motion can arise. In the example shown in fig. 40 a signal travelling at the speed of light from a point source strikes a rod made of fluorescent material, hitting the middle of the rod first. As the signal progressively strikes points on the rod, brief flashes of light are emitted. What the observer will see is a flash of light which

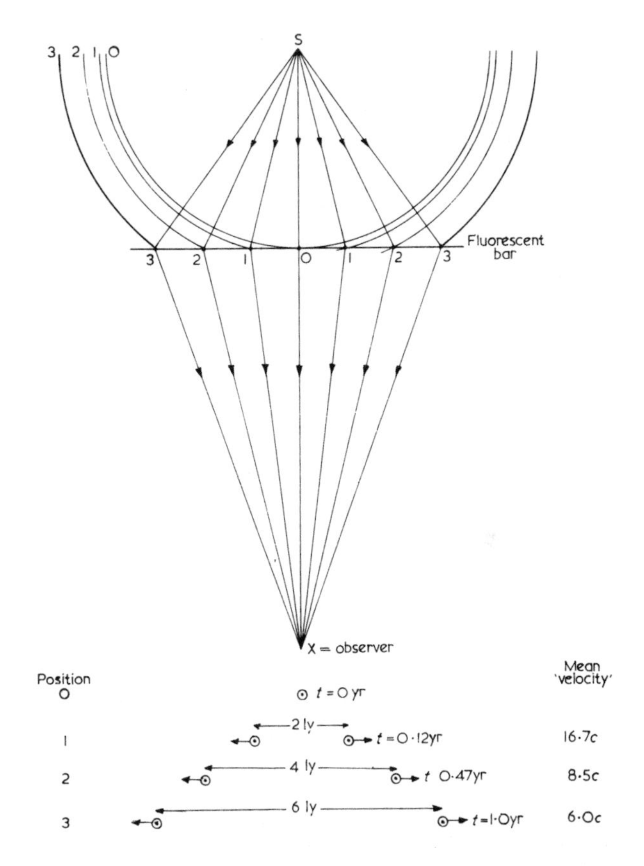

Fig. 40 *Apparent faster-than-light motion.* In this entirely hypothetical example we imagine a straight bar of fluorescent material, any point of which will emit a flash of light when struck by a photon of electromagnetic radiation. Imagine a pulse emitted symmetrically from the source S. This will strike the bar first of all at the centre (position 0) and then successively at points 1, 2, 3 along the bar on either side of 0. Light from the bar will reach observer X first of all from 0 and then from points 1, 2, and 3.

The fluorescent bar is assumed to be 6 light-years long and to be 4 light-years from S and 8 light-years from X; points 1, 2, and 3 lie at distances of 1, 2 and 3 light-years from 0. The flash from S will reach 0 after 4 years and X 8 years later. Due to the extra distance which it has to travel, the signal will not reach points 1 until 4.12 years after leaving S; i.e., points 1 will emit flashes of light 0.12 years after point 0. But points 1 are separated by 2 light years, and so X, after allowing for the fact that points 1 are slightly further from him than is point 0, will conclude that two patches of luminous material were "ejected" from 0 and separated from each other at an average speed of 2/0.12 times the speed of light—i.e., 16.7*c*. In fact, of course, there is no emission from 0; it is merely that signals from the common source S are reaching points along the bar at different times. As the signals "travel along" the bar, the rate of apparent separation diminishes.

divides into two points apparently travelling apart at a velocity well in excess of that of light. In fact the two points of light have not been ejected from the centre of the rod at speeds greater than light; it is merely that different parts of the rod are illuminated at different instants so that the geometry of the situation gives the impression of two moving objects.

Suitable geometric models involving jets, or ejected clouds moving out from a central source, can be used to explain the apparent faster-than-light separation of radio sources, and we can probably rest content that the speed-of-light barrier is still intact!

Seyfert galaxies are galaxies of a class first recognized in 1943 by the US astronomer Carl Seyfert. They appear to be spiral galaxies (our own Galaxy is a spiral) with extremely bright nuclei, their spectra showing the presence of excited gas moving at speeds of several thousand kilometres per second. Brighter than normal galaxies, in short-exposure photographs they look rather like stars because so much light is coming from the compact central nucleus. Most Seyferts are bright in the infrared part of the spectrum, but not particularly active at radio wavelengths. The central radio sources have proved to be too small to be resolved by existing techniques, and it has proved possible only to establish upper limits. The nucleus of the Seyfert NGC 4151 was photographed by a balloon-borne telescope and was shown to be less than 20 light-years in diameter—this is very much an upper limit, as brightness fluctuations indicate a central emitting region no more than a few light-years across and perhaps much smaller.

Quasars, Seyferts and the central components of strong radio galaxies display closely similar characteristics in having powerful, compact variable energy sources. The most "convenient" source available is an accreting black hole, and it is tempting to regard each entity as a different manifestation of the same basic kind of object—a galaxy with a violently active nucleus harbouring a central black hole which, because of its intense gravitational field, acts as the powerhouse responsible for the various observed phenomena. In a quasar the brilliance of the nucleus (the quasar proper) is assumed to be so great that the outer regions of the surrounding galaxy cannot be seen; in a Seyfert the central engine would be less powerful, so that the rest of the galaxy could be seen in longer-exposure photographs.

The Andromeda Galaxy, M31. Located at a distance of some 2,200,000 light years, it is the nearest large spiral galaxy, and contains over 100 billion stars. Although slightly larger than our own Galaxy, it is similar in appearance and general structure. Under good conditions it can be glimpsed without the aid of a telescope and is the most distant object which can be seen with the unaided eye. (Photograph from the Palomar Observatory, California Institute of Technology.)

The difference in luminosity of the nuclei would depend on a number of factors, notably the mass of the central hole and the amount of material available for consumption by it. A supermassive black hole into which gas and stars were pouring could be the most brilliant kind of object in the Universe. Assuming that an accreting black hole were able to convert infalling matter into energy with an efficiency of about 10%, the fuel requirement to keep a quasar shining would not be particularly excessive. An output typical of a quasar—10^{39} to 10^{40} watts—would require the digestion of no more than a few solar masses per year.

The assertion that Seyferts, radio galaxies and quasars are different aspects of the same kind of phenomenon would be more readily acceptable if we had direct evidence that quasars are indeed objects lurking in the central cores of galaxies, but so far it has not been possible to detect such associated galaxies. However, strong supporting evidence comes from the *BL Lacertae objects*. Named after BL Lacertae, which was at first classified as a variable star, and identified with an unusual radio source in 1968, they are brilliant point-like sources which exhibit rapid variability of greater degree and on shorter timescales than quasars themselves. Unfortunately, the spectra of these objects tend to be rather featureless, and do not exhibit the obvious emission lines which allow quasar red-shifts to be determined. Nevertheless, during the past few years, weak emission and absorption lines have been found in the spectra of some of them, and from these it has been possible to show that they have substantial red-shifts and are indeed contained within galaxies. In some cases, the surrounding galaxies can be seen directly, and it is now assumed that all BL Lacertae objects lie within galaxies.

The fact that BL Lacertae objects so closely resemble quasars in their properties, taken together with the fact that they are contained within galaxies, adds considerable weight to the assertion that quasars, too, lie in galactic nuclei. However, this has not yet been proved, and we should keep an open mind on the question.

It now looks as if the whole range of quasars, BL Lacertae objects, radio galaxies, Seyferts and various other peculiar and disturbed galaxies are indeed different manifestations of the same phenomenon, differing from each other only in degree. In each case the central energy source may be a supermassive black hole accreting material from its surroundings and releasing copious amounts of energy. Irregular accretion would lead to irregular outbursts, of the kind which may be responsible for the knots of intense radio emission found in some double radio sources, and radio jets.

Do these various types of object represent an evolutionary sequence, or are they distinct and separate classes of phenomena sharing the same fundamental mechanism? For example, it has been suggested that a quasar may evolve to become a radio or a Seyfert galaxy. Two extreme points of view are: (a) every galaxy was a quasar for part of its life; (b) a quasar remains a quasar for the entire lifetime of its parent galaxy. If proposition (a) is correct then, since quasars are very much rarer than ordinary galaxies, the active lives of quasars must be very short—no more than tens of thousands of years. On the other hand, if proposition (b) is correct, quasars would have to keep shining for more than 10 billion years, and we have severe problems trying to account for their energy outputs.

The present evidence favours the likelihood that the active period for quasars lies in the range from about 10 million years to a few hundred million years, and that the quasar stage is passed through by only a minority of galaxies. The quasar stage would end when the supply of infalling material dropped below a certain threshold, probably as a result of the quasar mopping up all the readily available matter in the galactic nucleus, and it is quite likely that "dead" quasars now outnumber "living" ones.

The presence of a lobe of radio-emitting material on each side of a central galaxy, and lying on a line through the centre of that galaxy, has for long been taken as evidence of the expulsion of material from galactic nuclei. In recent years very long baseline interferometry techniques have allowed radioastronomers to resolve similar pairs of sources on far smaller scales in the very cores of the central galaxies. The inner lobes, beams or jets are in some cases only a few light-years in extent, and *they are aligned in the same direction as the much larger outer radio lobes*. For example, the radio source 3C 111 has two outer lobes spanning 800,000 light-years aligned in the same direction as an inner pair of lobes separated by only about 3 light-years. The giant source NGC 6251 is particularly interesting: from the core of the central object emerges a radio-emitting jet some 6 light-years long lined up with a much longer jet some 720,000 light-years in extent, and both of these features are aligned with the outermost lobes, which span a period of 10 million light-years.

We see the same general pattern in a significant number of sources where central jets of material are aligned with clouds of radio-emitting material which lie perhaps millions of light years away and so must have been expelled from the nucleus at least several million years earlier in order to get out to these distances. How can the central source preserve the memory of the direction in which it ejected those earlier clouds in order, millions of years later, to expel fresh material in the same direction? A very massive source could remain stable against gravitational and other perturbations over such long periods, and the evidence strongly favours the assertion that this could be a spinning supermassive black hole, which would act like a giant gyroscope: jets of material squirted out along the line of its axis of rotation would thus all be pointed in the same direction.

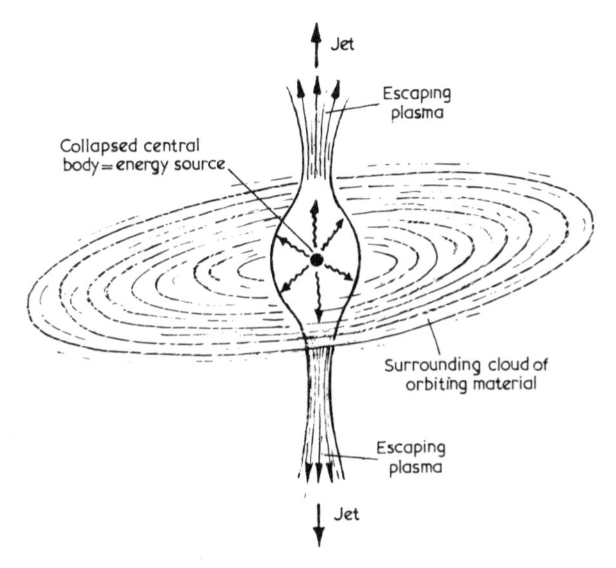

Fig. 41 *The Blandford-Rees model for twin-beam sources.* According to a model developed by R. D. Blandford and M. J. Rees, a cloud of hot energetic plasma (ionized gas), being energized by some central energy source (such as an accreting black hole), and contained by a circulating, flattened cloud of gas, will escape along the line of least resistance; i.e., it will be expelled through two "nozzles" lying along the perpendicular to the inner part of the circulating disc. This direction will coincide with the axis of rotation of the central black hole.

Various scientists have suggested models to account for material being ejected along the axes of rotating black holes. The model by

R. D. Blandford and M. Rees (fig. 41) considers the ultra-hot plasma which would be formed from the infalling material close to the black hole. The highly energetic particles produced there would escape most easily along the line of least resistance, perpendicular to the plane of the spinning cloud of infalling material; i.e., in the direction of the axis of rotation of the massive central black hole. In this way, streams of high-velocity particles of the type required for the radio-emitting lobes would be shot out in opposite directions along the axis of the hole. The source SS 433 referred to earlier (see page 167) may be an example of a double beam in action on a smaller scale.

Rees has suggested that the most violently variable quasars may be those whose "nozzles" point in our direction, so rendering readily visible any fluctuations in the jets. This may apply particularly to BL Lacertae objects. He argues that the emission lines in the spectra of ordinary quasars originate in a cloud of material surrounding the central source; but if the beam is directed towards us the emission-line radiation would be swamped by the beam, so accounting for the fact that with BL Lacertae objects we see little or no emission-line features.

It is interesting to note that Seyfert galaxies do not show the characteristic double-lobe appearance. Whereas double-lobe radio galaxies are usually found to coincide with elliptical galaxies, which contain very little gas, Seyferts look like ordinary spiral galaxies, which contain substantial quantities of gas. It seems reasonable to suppose that the reason double radio lobes are not seen with Seyferts is because the relatively large amount of gas present prevents the jets from emerging to large distances.

Various other mechanisms have been proposed to account for the double-lobe appearance. For example, M. J. Valtonen, of Alabama University, and his co-workers have proposed that mutual gravitational interactions between a number of black holes in a galactic nucleus could lead to "slingshot" ejection of black holes in opposite directions from the nucleus, the concentrated knots of emission present in some double sources being regarded as the sites in which these ejected holes are now located.

All in all, the evidence compellingly favours the proposition that supermassive black holes are the central "engines" which drive the whole range of active galaxies—Seyferts, BL Lacertae objects, quasars and strong radio galaxies—the masses required for these central holes ranging between 10 million and several billion solar masses.

However, other models have been proposed. For example, it has been suggested that the core of a quasar consists of a highly luminous supermassive star, or a concentrated cluster of massive stars where supernovae are exploding in a kind of chain reaction. Another suggestion is that what is involved is a collapsed mass which has not quite reached the black-hole state but is supported by intense circulating magnetic fields—a "magnetoid" or "spinar".

The problem that arises with most of these alternative objects is that they should in any case evolve rapidly into supermassive black holes. Certainly a supermassive star would do so in a very short time, and a spinar would lose energy and probably collapse to the black-hole state within at most a million years. The stars making up a massive star cluster should lose energy as they make mutual close encounters (if many of them have passed through the supernova stage they may anyway be black holes), as a result of which they eventually should fall together to generate a black hole. It would appear as if any massive body or collection of bodies at the centre of a galaxy should become a black hole eventually. Surely it is simpler to accept the black-hole model as the simplest and, at present, most plausible one. As Professor Martin Rees has put it, the supermassive black-hole model is the "best buy" at the moment.

There are numerous ways in which supermassive black holes could form in the centres of galaxies—for example, the direct collapse of a gas cloud; the collapse of a supermassive star which would evolve very rapidly (see page 110), its stellar phase representing a brief pause in the collapse of its gas cloud; the growth by accretion of a black hole formed by the collapse of a single massive star; or the growth by accretion of a black hole which formed before the galaxy which now surrounds it. In this last context, there is some evidence (see Chapter 11) to suggest that there may have been a burst of star formation at an early era before the galaxies settled into organized entities; indeed, it has been suggested that massive black holes may have provided the seeds around which the galaxies formed.

There is no shortage of hypotheses—proof is another question. There are several specific sources where evidence for a central black hole is persuasive.

The giant elliptical galaxy M87 in the constellation Virgo is a case in point. Short-exposure photographs reveal a narrow jet of light emanating from the nucleus, and radio observations indicate that M87 is a powerful emitter which may also be a source of short-lived radio bursts. It is, too, an X-ray source. In 1978 W. L. W. Sargent and P. J. Young of the Hale Observatories, A. Boksenberg and K. Shortridge of University College, London, C. R. Lynds of Kitt Peak National Observatory, and F. D. A. Hartwick of the University of Victoria published spectroscopic results which showed that stars less than 300 light-years from the galaxy's centre were moving dramatically faster than those further out. To account for this, there has to be a very large mass concentrated in the core of the galaxy. The results were consistent with there being a central black hole of about 5 billion solar masses.

A parallel investigation by P. J. Young, J. A. Westphal, J. Kristian and C. P. Wilson of the Hale Observatories and F. P. Landauer of NASA's Jet Propulsion Laboratory examined the surface brightness of the galaxy and found an intense point-like source at the centre. The luminosity and colour of this intensity "spike" cannot be explained by a dense concentration of ordinary stars, and the observations are again consistent with the presence of a black hole of some 5 billion solar masses.

M87 is not a brilliant source like a quasar, and is virtually devoid of gas. Calculations indicate that a mass flow into the hole of only 1% of the Sun's mass per annum, converted into energy at only 0.002% efficiency, would be sufficient to account for M87's output, and so the central black hole could readily be fuelled by mass loss from stars in its vicinity.

One possibility is that M87 represents a dead quasar. In the distant past, when more gas was available as fuel for the central hole, M87 could easily have matched the most brilliant quasar.

Another intriguing source which has aroused a lot of interest is the radio galaxy Centaurus A. At a distance of only 16 million light years, it is the nearest radio galaxy and, although not a particularly powerful source, it displays the characteristic double-lobe structure. Optically it is a strikingly beautiful object, looking like an elliptical galaxy crossed by a dense band of dust. At X-ray wavelengths, it has been found to have a jet which aligns with similar radio and optical features. Although the radio output (10^{35} watts) is tiny compared to that of a quasar, it is nonetheless significant. Within the core of the galaxy there lurks a tiny variable source of radio, infrared and X-ray emissions—observations have revealed that the radio-emitting region is less than a light-day across, and the X-ray source no more than a few light-hours in diameter. Although the radio power is low now, the amount of energy contained in the particles making up the extended radio lobes indicates that Centaurus A must have been far more active in the past, and it may be that in its core there resides a black hole of 10 million solar masses or more, now largely starved of fuel. If these ideas are correct, Centaurus A could contain the nearest actively accreting supermassive black hole.

In May 1979 D. Walsh of Jodrell Bank, England, R. F. Caswell of Cambridge and R. J. Weymann of Steward Observatory, Arizona, reported observations of what appeared to be a double quasar, a source (known as 0957 +561 A,B) consisting of two virtually identical quasars of very similar brightnesses, identical red-shifts, and separated by an angle of only 5.7 seconds of arc. The distance indicated by the red-shifts would imply that the quasars are separated by only some 220,000 light-years. Quasars are rare objects, and to find two with identical properties so close together is highly improbable.

The alternative which the authors proposed was that there is only one quasar, but that between it and the Earth lies a body of perhaps 10^{13} solar masses acting as a gravitational lens. The slight difference in brightness and in the radio structure of the two components could be accounted for if the "lens" were slightly to one side of the direct line of sight. Different groups of radioastronomers disagreed over the detailed radio structure of the source (although an optimistic eye could discern a resemblance to the twin crescent shapes that would be expected to characterize a source seen through an off-axis lens), but optical results of observations made later in 1979 by groups at the Steward and Hale Observatories seem to have shown fairly conclusively that a galaxy lies in the foreground (having a red-shift only one third that of the quasar) between the two observed components.

It seems reasonable to conclude that it is this galaxy which is acting as a lens, rather than a black hole as such. Nevertheless, if these results are confirmed they provide a striking vindication of the predictions of General Relativity.

★ ★ ★

The centre of our Galaxy is hidden from the optical astronomer's gaze by dense clouds of dust which are so effective at obscuring light that only about one photon in every ten billion or so actually survives its 30,000-light-year journey to Earth. Fortunately, radiations of other wavelengths can penetrate the dust, and the galactic nucleus has been shown to be a source of radio, infrared and X-ray emissions.

At the galactic centre lies a radio point source known as Sagittarius A (west) which has been shown to be smaller than 10 astronomical units across—i.e., the radio-emitting core of the Galaxy is smaller than the orbit of Saturn. Some of the infrared sources turn out to be cool red giants but one, which coincides with Sagittarius A (west), cannot be explained in this way. Study of microwave emission lines (of the element neon) in gas clouds at the galactic centre show high velocities which can best be explained by assuming the clouds to be moving in the gravitational field of a mass of 5 to 8 million solar masses concentrated within the central parsec (i.e., within a region roughly 3 light-years in diameter). Although stars are expected to be concentrated more closely in the nucleus than elsewhere, infrared observations indicate that the total of the stars in the central parsec is only about 2 million solar masses.

There thus seems to be an additional mass of between 3 and 6 million solar masses concentrated at the heart of the Galaxy, and a plausible explanation is that it is contained within a massive black hole which provides the underlying powerhouse for the compact radio, infrared and X-ray source. There is evidence, too, for violent events in the past history of our Galaxy, although not on the scale associated with quasars or radio galaxies: observations reveal clouds of gas sweeping out from the galactic nucleus. V. M. Clube of the Royal Observatory, Edinburgh, argues that the central object is a spinar (see page 183) which periodically explodes and reforms, giving rise to bouts of ejection of material from the nucleus; in his view "new physics" will be required to explain this process. The more conventional* view is that a massive black hole is itself quite capable of explaining the observed phenomena. So our Galaxy, too, may harbour a black hole in its core, and may have periodic bouts of activity whenever the accretion rate increases.

*It is interesting to note how, within a few years, black hole models of galactic nuclei have come to be regarded as the *conventional* explanation!

Globular clusters are massive, ancient star clusters which contain from a few tens of thousands up to about a million stars concentrated in a relatively small volume of space; over a hundred are known to be associated with our Galaxy. Over a long period of time, the most massive stars in such clusters are expected to settle towards the centre, and it is possible that they may coalesce to form black holes. It has been suggested that accretion onto black holes of up to a few thousand solar masses might be responsible for the observed X-ray emission from some globular clusters, several of which show central intensity "spikes" rather like that found in M87 (see page 184).

It is puzzling, however, that the X-ray sources in clusters tend not to be significantly brighter than those associated with binaries where there is accretion onto a neutron star. Many globular clusters contain curious X-ray sources known as "bursters" which switch on and off erratically, flaring up and down again in a matter of seconds. At the time of writing two main theories concerning these hold the centre stage. On the one hand we may be witnessing X-ray surges due to erratic accretion from companion stars onto the surfaces of neutron stars (J. Grindley, team leader examining observations of such events made by the Einstein Observatory, has described the infalling material as "helium bombs"); on the other, the X-rays may come from violent instabilities in matter circulating close to the event horizons of massive black holes.

There is little to choose between these possibilities at the moment, although the presence of luminosity "spikes" at the centres of some globular clusters adds circumstantial evidence favouring the accreting-black-hole model. Continuing X-ray observations, and data from the forthcoming gamma-ray satellite, may resolve the issue in the near future.

Mini black holes

To date there is no evidence to suggest that mini black holes exist. If such holes do exist, and it is a big "if", they must have formed in the first instants of the Universe, and should more properly be referred to as primordial black holes. As we saw in Chapter 9, primordial holes of about a billion tonnes should be exploding now, and the best prospect of detecting them might be to look for the bursts of energetic gamma rays released in the explosions.

The chances of detection with existing gamma-ray detectors are very slim, and even on the most optimistic estimates it is unlikely that existing or immediately projected detectors could pick up an individual explosion at a range greater than a fraction of a light-year.

All that can be done is to set upper limits. If it is assumed that *all* of the existing gamma-ray background in the Universe was produced by exploding mini holes, calculations made both by D. N. Page and S. W. Hawking and by G. F. Chapline imply that there cannot be more than about 300 million primordial black holes per cubic light-year if the mini holes are concentrated in galaxies. If the holes are spread uniformly throughout the Universe, the value drops to only some 300 per cubic light-year, *as an upper limit.* There may be fewer (indeed, there may be none at all).

Another possibility is to use the Earth's atmosphere as a gamma-ray detector. A high-energy gamma ray hitting the atmosphere produces radiation which in principle could be detected from ground level as a flash of light. Experimental results by N. A. Porter and T. C. Weekes (University College, Dublin) indicate that at most there are 2 black-hole explosions per century per cubic light-year in our part of the Galaxy. D. Fegan and S. Danaher (also of University College, Dublin) have proposed using the 5,500 mirrors of the world's largest solar energy plant at the Sandia Laboratories, New Mexico, to hunt at night for flashes of light which may be due to exploding primordial black holes. At the time of writing, there were no results to report.

A more promising approach was proposed in 1977 by Rees, who argued that particles released by black-hole explosions would interact with the galactic magnetic field to produce a linearly polarized pulse of radio waves which could be detected with existing techniques much more easily than could a gamma-ray pulse. Under optimum conditions, existing radiotelescopes should easily be able to detect such bursts as far away as the galactic centre, and the largest single radio dish, at Arecibo, Puerto Rico, should in principle be able to detect a single explosion as far away as the Andromeda Galaxy! There are many assumptions involved here, but this seems a promising line of approach. Preliminary analyses of radio surveys discussed in 1977 by W. P. S. Meikle indicate that, at most, there is one primordial-black-hole explosion per cubic light-year per three million years. This limit is about a hundred thousand times more sensitive than the gamma-ray detection limit.

Finally, what about the chances of a direct encounter? Is there any possibility of the Earth being struck by a mini black hole? In 1973 A. A. Jackson and M. P. Ryan of the University of Texas published a remarkable paper suggesting that the impact of a mini black hole could have been responsible for the "Tunguska event" which took place in Siberia in 1908. As a result of some kind of explosion, trees were flattened over an area of thousands of square kilometres: at first the explosion was ascribed to the impact of a large meteorite, but the absence of a crater seems to preclude this. The mini hole postulated by Jackson and Ryan was considered to have been of a mass comparable with a small asteroid; even so, it would have been less than a millionth of a centimetre in radius, and would have passed straight through the Earth and out the other side. Considerable quantities of energy, they argued, would have been released as the hole penetrated the atmosphere and hit the surface—perhaps as much as a 20-megaton hydrogen bomb, and this would have devastated the surrounding forest. Alternative "explanations" of this event have included the explosion of an alien spaceship, but the most reasonable suggestion is that it was the nucleus of a small comet which struck the Earth, exploding in the atmosphere and so not producing a large and obvious crater. Jackson and Ryan's mini-hole theory—although not taken seriously—was at least more plausible than some.

In point of fact, the chances of the Earth being struck by a mini black hole are infinitesimal. According to Bernard Carr of the Institute of Astronomy at Cambridge, any hole less massive than about 10^{21}kg is unlikely to produce readily observable effects on passing through the Earth, and even on the most optimistic estimates we should expect only one such encounter in a period of time a thousand times longer than the estimated age of the Universe. Even for the lower mass considered by Jackson and Ryan we could expect no more than one collision in 10 billion years. On the face of it, direct encounters do not hold out any promise of detecting black holes, although this argument does not rule out the possibility that there might exist primordial black holes, of mass appreciably greater than a billion tonnes, somewhere in the Solar System.

We have at present, then, no evidence at all for the existence of mini black holes. However, we have good reason to suppose that we are seeing evidence of the existence of stellar-mass black holes in close binary systems, and of supermassive black holes in galactic nuclei and quasars. There is mounting evidence in favour of massive black holes being the underlying energy sources in a wide variety of violent and energetic phenomena ranging from the modest activity of our own galactic nucleus to Seyfert galaxies, radio galaxies, BL Lacertae objects and quasars. Gravity may, therefore, be the force responsible for the most energetic entities in the known Universe.

11
Gravity and the Universe

Gravity controls the motion of all the bodies in the Universe, determines the shape of space itself, and governs the evolution and ultimate fate of the Universe as a whole.

In order to understand the structure and evolution of the Universe we have to collect every scrap of information which comes our way, and our ability to do this is very much controlled by the instruments and techniques available—in this respect the present-day astronomer has immense advantages over his predecessors of only a couple of decades ago. The Universe is a vast place; we are part of it and we cannot get "outside" to take a detached view of its structure. We cannot see the whole of it, and we do not know what may lie beyond the range of our telescopes. Because light travels at a finite speed, the information which we are now receiving from great distances is very "out of date"; there is no way in which we can see the Universe "as it is now" (indeed, as we have seen, there is no way that we can assign absolute times to events in the Universe). On the other hand, the fact that our information often relates to times billions of years in the past allows us to study directly what conditions were like much earlier in the history of the Universe.

It is an understatement to say that attempting to understand the Universe as a whole is a gargantuan task. Nevertheless, during the past two decades one particular theory of its origin and evolution has gained wide acceptance; and cosmology has moved from an era when there were many theories but scarcely any observations to one in which observational data exists in quantity, and it is possible to apply critical tests capable of disproving many otherwise plausible models of the Universe. Such is the confidence of some theorists at least that in their view the entire history of the Universe can be charted back to the first hundredth or even the first millionth of a second of its existence.

The Universe about us appears to have various hierarchies in its structure. Planets revolve around stars, stars are gathered together in galaxies, galaxies themselves are members of clusters of galaxies, and there is evidence to suggest that the clusters themselves may be gathered together in superclusters. There may even be supersuper-clusters! Astronomers have attempted to establish the structure of the Universe by determining the motions and distributions of galaxies, radio galaxies, quasars, clusters and the like—the *visible* matter in the Universe. However, evidence is mounting to suggest that there is more to the Universe than the eye can see, that visible material may be far outweighed by invisible material which could, by its gravitational influence, decide the ultimate fate of the Universe.

Central to any attempt to determine the structure and evolution of the Universe is the question of distance measurement, and it is in this vital area that considerable scope for error exists, the magnitudes of the errors rising rapidly with increasing distance.

There are three basic approaches to distance measurement. The first and most common method essentially compares the measured apparent brightness of a source with the luminosity which we would expect that source to have; thus, if we know the luminosity which a particular *type* of star has, we can measure the amount of light received from any star of that type and calculate its approximate distance. This technique, known as the *distance-modulus* method, relies upon our being able to identify in distant galaxies particular kinds of highly luminous objects, and upon our ability to make accurate estimates of their luminosities. With greater distances the properties of entire galaxies, and the statistical properties of clusters of galaxies, must be used, and it is clear that the errors in the various links in the chain of reasoning grow and compound each other.

The second approach relies upon identifying objects of known size (or, at least, of a type whose size can be estimated)—objects within galaxies, or galaxies themselves—and comparing their angular sizes with their assumed linear sizes. This technique is even more fraught with potential for gross errors of observation or interpretation. A third technique depends upon the red-shifts in the spectra of galaxies and quasars, of which more will be said shortly.

The distance to another galaxy was first determined in 1923. In that year E. P. Hubble (1889–1953) used the 100in (2.5m) reflector at Mount Wilson to study stars known as Cepheid variables in the Andromeda Galaxy, M31, these stars being highly luminous variable stars having the convenient characteristic that their periods of variation are related to their luminosities (the more luminous the Cepheid, the longer its period) according to a law established a few years earlier by Henrietta Leavitt (1868–1921). By measuring their periods it was possible for Hubble to work out their luminosities and, by comparing these with their observed brightnesses, to establish the distance of the galaxy in which they lay.

Hubble's work suggested the Andromeda Galaxy lay at a distance of about 750,000 light-years. Subsequent revisions have increased that figure by a factor of three.

The Universe—some basic questions and a fundamental observation

How is the Universe arranged? How does it behave and evolve? Is it finite or infinite in extent? Did the Universe originate a finite time ago, or has it always existed? Will it continue to exist forever, or will it eventually come to an end?

These are some of the key questions which give cosmology its inherent fascination. Today, we may be close to answering some of them.

To Newton, the Universe had to be infinite. His theory of gravity, which coped very well with the motions of stars and planets, ran foul of one major difficulty when the Universe as a whole was considered. If the stellar Universe were finite in extent, with gravity acting mutually between every particle of matter, why did it not collapse into one mass? A stationary distribution of stars, however extensive, would fall together under gravity until all the masses were concentrated together at one point. To overcome this difficulty Newton postulated that the Universe was infinitely large so that the net attraction on any given point would "cancel out" and there would be no unique centre towards which everything would fall.

In this context there is one fundamental observation: *the sky is dark at night*. Why should this be so? The answer is by no means as obvious as it seems. Posed in 1826 by the German mathematician H. W. Olbers (1758–1840), the problem is generally known as

Olber's Paradox, although it had been discussed earlier by Halley and by J. P. Loys de Chéseaux (1718–1751). The argument goes as follows: if the Universe is infinite in extent and uniformly filled with stars then, in whichever direction you look, eventually you should be looking at the surface of a star (although we know that stars are clumped together in galaxies, a similar argument can be framed in terms of galaxies), and so the whole sky should be at least as bright

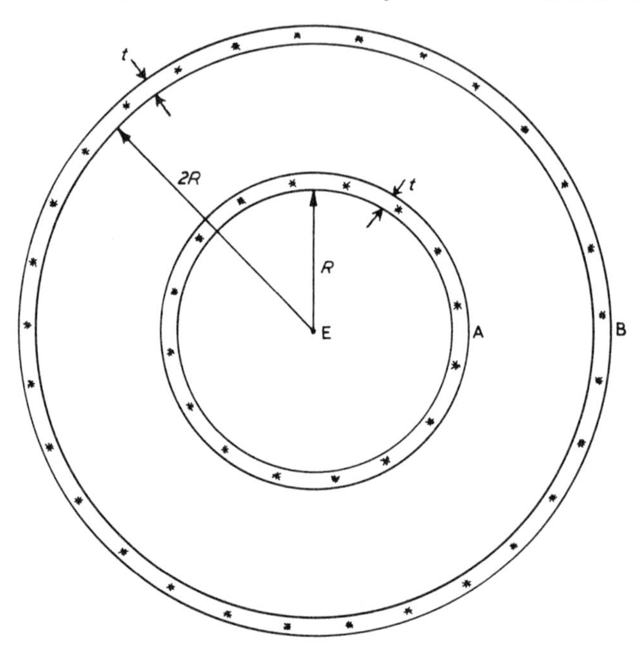

Fig. 42 *Olbers' Paradox.* Imagine the Universe to be uniformly filled with stars and to be of infinite extent. Consider two thin spherical shells drawn in space and centred on the Earth, shell B having twice the radius of shell A, and both shells having a thickness t. For simplicity we will assume that all stars emit equal quantities of light. A star in shell B is twice as far away as a star in shell A and, because the brightness of a source diminishes with the square of its distance, we on Earth (E) will receive only one quarter as much light from it. The number of stars in a shell depends on its surface area multiplied by its thickness (t). The area of a shell depends on the square of its radius: shell B therefore has an area four times as great as that of A, and so contains four times more stars than does shell A. The two effects therefore cancel out, and the amount of light received at E from each shell is exactly the same.

If the Universe were of infinite extent, an infinite number of shells could be drawn. Clearly foreground stars would block out some of the background stars, but the net result should be that the whole sky ought to be as bright as the surface of the Sun. (In fact, due to the "blocking out" effect of foreground stars, the Universe would need to be "only" 10^{26} light-years in radius for the whole sky to be as bright as the Sun.)

Star field in the direction of the galactic centre. Looking towards the centre of the Galaxy, in the constellation of Sagittarius, vast numbers of stars may be seen. However, because of the obscuring effect of interstellar dust, the centre itself cannot be seen in visible light. (Photograph from the UK Schmidt Telescope Unit, Courtesy Photolabs, Royal Observatory, Edinburgh.)

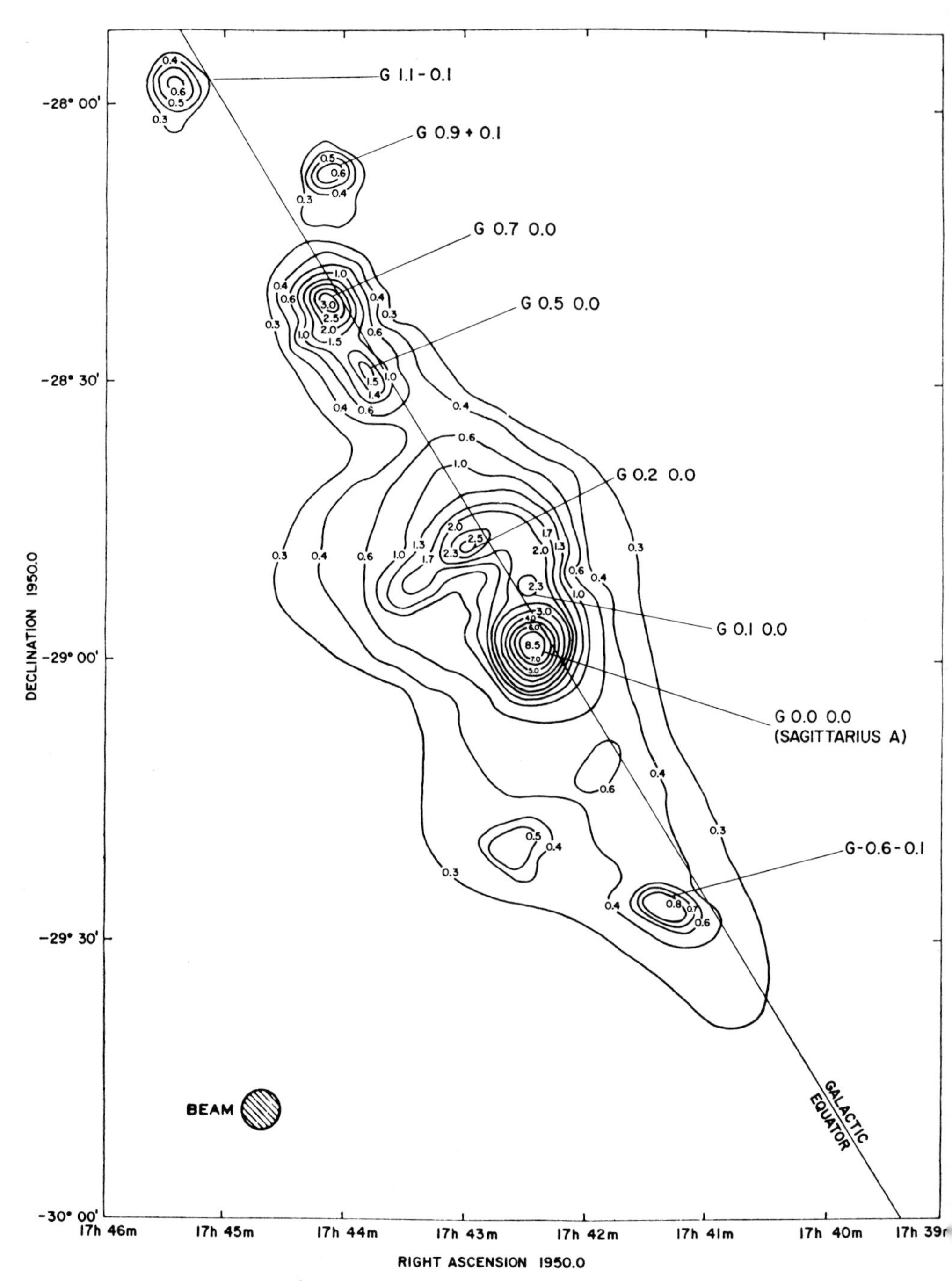

Radio map of the galactic centre, made at a wavelength of 3.75 centimetres by Dennis Downes and Alan Maxwell of Harvard University. The principal source of radio emission is ionised gas. The source Sagittarius A marks the centre of the galaxy, much of the radiation coming from a region of space smaller than the Solar System. There is considerable debate concerning the nature of the central energy source; there is a fair body of evidence to support the assertion that the galactic centre may harbour a black hole of several million solar masses. (Map courtesy of Dr Dennis Downes.)

as the surface of the Sun. The argument is examined in more detail in fig. 42.

What can we deduce from the fact that the sky *is* dark at night? The Universe could still be infinite and filled with stars provided that the stars have not been shining throughout the infinite past. If they began to shine a finite time ago there would not have been sufficient time for light to reach us from the more distant ones, and the problem would be circumvented. On the other hand, it is difficult to see how all the stars in an infinite Universe should have "switched on" at the same time. The conclusion apparently to be drawn is that the Universe cannot consist of a stationary distribution of stars which is infinite in age and extent. The darkness of the sky places powerful restrictions on any theory of cosmology, and makes Newton's view of the Universe seem highly implausible.

General Relativity and cosmology

General Relativity appeared on the scene at a time when the Universe was still considered to be static. When Einstein himself first attempted to apply General Relativity to the Universe in 1917 he found that it was necessary to add an extra term to his equations in order to permit it to be static. In effect, this "cosmological term" was a cosmic repulsion which held the masses apart to prevent their coming together, and its introduction amounted to "fiddling the books" to get the "right" answer: a static Universe.

Two years later the Dutch physicist W. de Sitter (1872–1934), with whom Einstein had copious correspondence, obtained a solution of the field equations for a universe devoid of matter (hardly a realistic situation) which had the property that, if test particles were placed in it, they would fly away from each other with ever-increasing velocity. Then, in 1922, the Russian mathematician A. Friedmann (1888–1925) obtained solutions for expanding matter-filled universes. Three basic Friedmann models still provide the basis of most current work in cosmology: according to these models the Universe originated in a singularity, and can be "open", "flat" or "closed". The open Universe is one which will expand without limit, and where space and time are unlimited in extent. The closed Universe is of finite extent in space and time, expands to a finite size, and recollapses into a singularity. Between these two possibilities lies the flat

Universe, which also expands without limit but where the expansion rate of its constituent parts tends towards zero in the infinite future. As we shall see shortly, twentieth-century observations are consistent with Friedmann models, and imply that the Universe is expanding from an initial singularity.

Virtually all theoretical cosmological models incorporate the *Cosmological Principle*, which states that the Universe is *homogeneous* and *isotropic*. By "homogeneous" we mean that, wherever you are located in the Universe, its large-scale appearance will be the same; and by "isotropic" we mean that the Universe looks the same in every direction. All the observational evidence suggests that these conditions apply. In addition we have to assume that the properties of matter and radiation, at the basic laws of nature, are everywhere the same; again, all the evidence to date favours that hypothesis. If we could not make these assumptions, it would be virtually impossible to construct a model of the overall structure and behaviour of the Universe.

Cosmological models treat the Universe as being of uniform density throughout; i.e., as if all the matter of the stars and galaxies were smeared out rather than being concentrated in lumps. Taken on the large scale, this is a good approximation. Indeed, the observed isotropy and uniformity of the Universe poses problems for theoreticians wishing to account for the existence of galaxies; for how can the Universe be so "bland" yet still have given rise to the density fluctuations which correspond to galaxies and clusters?

The Hubble law

Following the discovery by M. Humason (1891–1972) that a number of galaxies showed red-shifts in their spectra which—interpreted in terms of the Doppler effect—implied that they were receding from us, Hubble investigated more galaxies and found by 1929 that their red-shifts were proportional to their distances. Subsequent investigations to greater and greater distances have shown the validity of the *Hubble law*, which states that all the galaxies (beyond our immediate neighbours which make up the Local Group) are moving away from us with velocities directly proportional to their distances (fig. 43). We can write this law as $V = H_0 \times D$, where H_0 is the present value of a number known as *Hubble's constant*. Velocity in this case is usually expressed in km/sec and distance in megaparsecs; 1 megaparsec (Mpc) is 1 million parsecs,* or about 3,260,000 light-years. Hubble's constant is notoriously difficult to measure, different methods of determination yielding different results. Values quoted in the current literature range from about 40km/sec/Mpc to about 110km/sec/Mpc, but the most commonly quoted "consensus" value is 55km/sec/Mpc. This implies that a galaxy at a distance of 1 megaparsec is receding from us at a speed of 55km/sec; by Hubble's law, a galaxy ten times further away will be receding at 550km/sec, and so on.

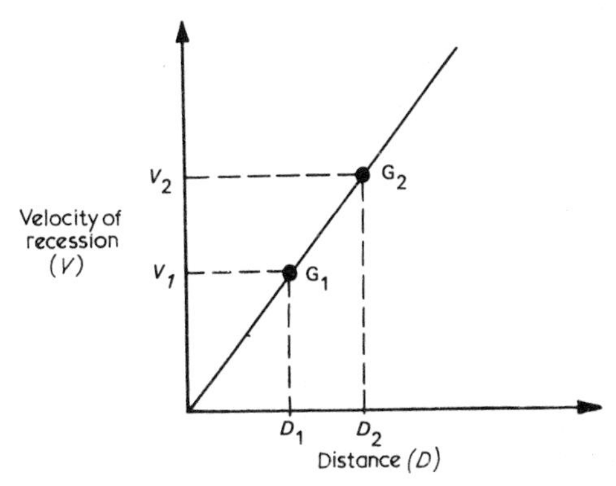

Fig. 43 *The Hubble law.* The velocities at which galaxies are receding are directly proportional to their distances. For example, galaxy G_2 is at twice the distance of galaxy G_1, and is receding at twice the velocity. The slope of the line gives the relationship between velocity and distance and is equal to the value of Hubble's constant: i.e., $V/D = H_0$.

This striking and fundamental observational law lies at the heart of our present view of the Universe. Hubble's observations showed that the Universe as a whole is expanding, and that there was no need for a "cosmological term" to be added to Einstein's equations to keep the galaxies from falling together.

The expansion of the Universe eliminates Olbers' Paradox. The more distant the galaxy, the faster it is receding, and the greater the red-shift in its spectrum. The effect of a strong red-shift is to weaken the radiation from a source: beyond a certain distance the red-shift is so great as to prevent our seeing anything. The Hubble law—simply interpreted—implies that there is a boundary at least to the *observable*

*A *parsec* is a unit of distance based on the parallax method used by astronomers to establish relatively small stellar distances: 1 parsec = 3.26 light-years.

Universe. In the language of General Relativity we would say that the red-shift imposes a "horizon" beyond which we cannot see. If we cannot receive light from objects beyond a given distance there is no problem about our having a sky which is dark at night.

Does the Universe have a centre?

At first glance the Hubble law appears to indicate that we are in a privileged position—at the centre of the universal expansion. Everything else in the Universe appears to be moving away from us. We should be very suspicious of any observation which suggests that we occupy a unique position in the scheme of things: every time we have set ourselves up as being in a position of particular significance we have been forced to admit the error of our ways. In any case, if the Earth were at the centre of the Universe, this would violate the Cosmological Principle, for the Universe could not possibly look the same from every point within it.

In fact the observations indicate that the entire Universe is expanding, that each galaxy is receding from every other galaxy. The scale of the Universe is increasing, the separation between points within it is becoming greater, but no one galaxy can claim to lie at the true centre of this expansion. No matter on which galaxy you were located you would see the same general picture—every other galaxy would appear to be receding from you in accordance with the Hubble law. It is easy to visualize this for a string of galaxies lying in a straight line and separating from each other with velocities proportional to distance (fig. 44), although, obviously, we find it less easy to visualize in three dimensions.

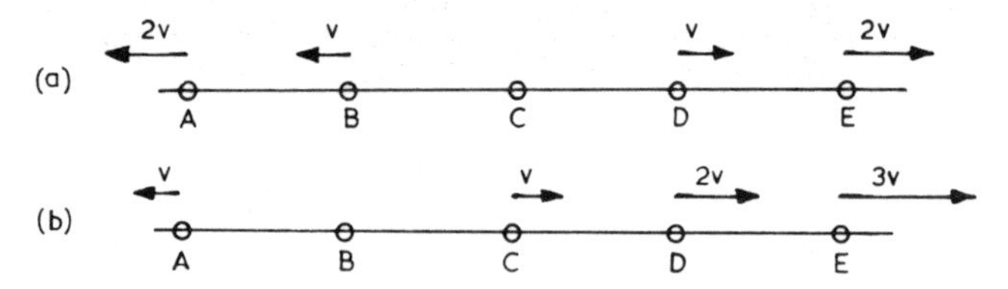

Fig. 44 *Relative velocities of receding galaxies.* Imagine a string of galaxies evenly spaced out along a line and separating in accordance with Hubble's law. (*a*) From the point of view of an observer on galaxy C, B and D are receding from him in opposite directions at velocity v. (*b*) So far as an observer on galaxy B is concerned, C is receding from *him* at velocity v, A is receding in the opposite direction at speed v, D is receding at velocity $2v$, and so on. Each has the impression that he is at the centre of the expansion.

The universe may not have a "centre"; and it may not have an "edge" either. General Relativity allows the possibility of a Universe which is finite yet unbounded; i.e., a Universe which contains a finite volume of space yet which has no discernible edge. The net effect of all the matter in the Universe may be such as to curve space (or, strictly, space-time) into a closed system so that, in principle, a ray of light setting off in one direction eventually would return to its starting point without ever having encountered an edge to the Universe.* By far the best analogy for the closed Universe is to think of it as being the *surface* of a balloon. We know that space has three dimensions (along, across and up), but we have to imagine all of space mapped onto the surface of the balloon, and we have to accept that the inside and outside of the balloon have no significance for us—we cannot leave the surface of the balloon, or penetrate inside it.

Imagine, now, two flat, two-dimensional creatures who live on the surface of this spherical balloon (we met such creatures in Chapter 5—see page 90). Being two-dimensional, these creatures could have no conception of the vertical direction. If one were an experimental physicist and the other a philosopher, and they made measurements of a small area of the sphere, they would conclude that their world was flat. They might debate the question of whether or not their universe was finite: how big was it? did it have an edge? If, leaving the philosopher to contemplate the problem, the physicist were to decide to attempt to find the edge of his universe, he might choose to set out in a given direction and keep going until he reached the edge. If he tried this experiment he would find after a time that he had returned to his starting point, probably to the considerable surprise of both himself and his philosopher friend. They would conclude that their universe was finite yet unbounded and, although they could not hope to *visualize* what a sphere looks like, nevertheless they could carry out measurements on the surface which would allow them to discover that their universe had the geometry of an abstract mathematical concept—which we, as external three-dimensional creatures, call a sphere.

Our own situation may be similar, in that we may live in a

*Although in fact the expansion and red-shift would prevent a would-be circumnavigating photon from ever being seen back at its source.

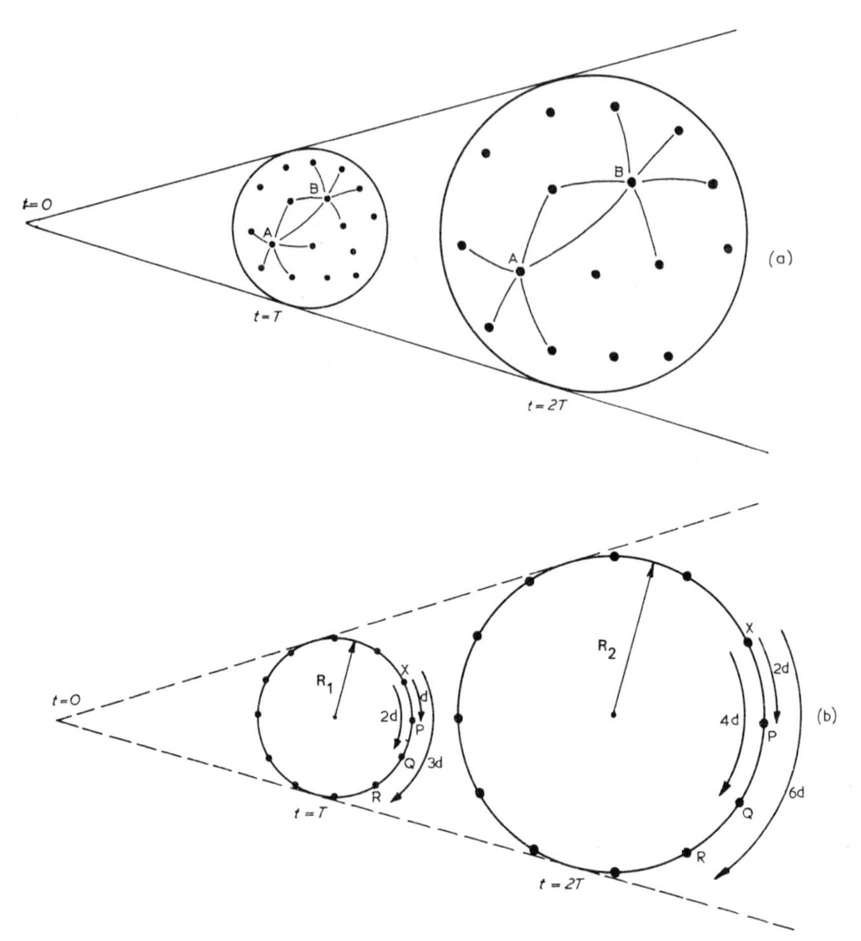

Fig. 45 *The expanding Universe—a balloon analogy*. Imagine the galaxies to be dots on the surface of an expanding balloon. (*a*) As the balloon doubles in size between times $t=T$ and $t=2T$ the pattern of dots remains the same, but the distances between them increase by a factor of 2. An observer on galaxy A observes that all the other galaxies are receding from him; an observer on galaxy B, or any other galaxy, will see the same picture. None can claim to be at the centre of the expansion, although if we look back to time 0, when the balloon had zero radius, then all the "galaxies" were at the same point at that time. (*b*) Looking at the balloon in cross-section we see how the scale of the balloon, set by its radius, changes between times $t=T$ and $t=2T$. R_2 is twice the size of R_1. If we consider a chain of evenly spaced galaxies around the perimeter of the balloon we can see how the Hubble law is obeyed. From the point of view of X, the distance of galaxy P increases from d to $2d$; i.e., by one unit of distance, d. Its "velocity of recession" is the change in distance divided by the time which has elapsed; i.e. d/T. Galaxy Q has moved from $2d$ to $4d$, increasing its distance by $2d$. Its "velocity of recession" is $2d/T$; i.e., twice as great as that of galaxy P. Likewise, the "velocity" of R is three times greater than that of P.

Universe with a closed geometry. We cannot visualize the shape of our Universe, but we can make observations of distant galaxies and other phenomena which may allow us eventually to determine the geometry of our Universe, whether it be open, flat or closed. For the moment, though, our observational data is insufficient to resolve the question.

Let us represent our Universe by the surface of a spherical balloon, and let us show the galaxies as dots on the surface (fig. 45). As we blow up the balloon, the separation between dots (galaxies) increases in proportion to the radius, or "scale factor", of the balloon, and each dot "sees" the others receding with a velocity proportional to distance. No one dot can claim to be the centre of such a universe, for each observer sees the same general picture in accordance with the Cosmological Principle. Tracing back the history of the expanding balloon shows that at some past time its radius would have been zero, and every point on its surface would coalesce into one and the same point (this isn't true of real balloons, of course, but it seems to be true of real universes). In a sense, every point on the surface can lay an equal claim to *having been* at the centre!

If we assume that the Universe had a singularity in its past—i.e., that everything in it originated a finite time ago in what has been termed the Big Bang, and has been rushing apart ever since as the result of this violent initial event—then the balloon analogy helps to make clear a crucial aspect of General Relativity's picture of the Big-Bang Universe. The material content of the Universe did not erupt forth from a hyperdense "primeval atom" (as the Abbé Lemaître (1894–1966) suggested in the early nineteen-thirties) into a preexisting space-time; matter did not suddenly pour forth into a previously empty space. Instead, *space and time originated* with the Big Bang. Neither time nor space, in the sense in which we use these terms, existed "before" that initial event. The question "What happened before the Big Bang?" has no meaning.

According to this viewpoint, galaxies are not rushing apart "through" space. Like the dots on the surface of the balloon, they are being carried apart by the expansion of space itself; indeed, we can visualize the galaxies as being at rest in an expanding space. We can regard the "smeared-out" average of all the matter in the Universe to be expanding like the surface of the balloon; and we say that an observer is a "fundamental observer" if he is at rest with respect to the "smeared-out" matter in his locality. To such observers the Universe appears homogeneous and isotropic. Apart from minor motions peculiar to the galaxies concerned, such as motion within the Local Group due to the gravitational influence of the galaxies in our immediate vicinity, an observer residing in any galaxy which is sharing in the general expansion of the Universe can claim to be a fundamental observer. The universe will appear the same to all such observers, and will be seen by them to pass through the same sequence of evolutionary stages; because of this it is possible to set up a general timescale applicable to events involving the Universe and on which all fundamental observers will agree—*cosmic time.*

Motion relative to the fundamental observers in their locality gives rise to length contraction and time dilation effects, such that if you were to set off from Earth in a spacecraft travelling at a large fraction of the speed of light you would not see an isotropic Universe; for example, some of the galaxies in your forward hemisphere of vision would be seen to be blue-shifted.

The age of the Universe

If our simple Big-Bang picture is correct, it should be possible, knowing the speeds of the galaxies and their distances, to work out when they, together with our own Galaxy, were packed together in the initial singularity.* The simplest approach is to assume that the galaxies have been moving away from each other with constant velocities since the initial event (we can visualize the more distant galaxies as having attained their greater distances because of their greater velocities). For any galaxy at distance D from our Galaxy, moving with velocity V, the time taken to reach its present distance is simply D divided by V: D/V. By Hubble's law the velocity and distance of any galaxy are related by $V=H_0 \times D$, and from this we see that the time required for any galaxy to reach its present distance is $D/V=1/H_0$. All the galaxies must have taken precisely the same time to reach their present distances, the more distant ones moving faster, according to the Hubble law, in just the right proportion to achieve this. An observer in any other galaxy would come to the

*Of course, the galaxies could not have assumed their present forms until some considerable time after the Big Bang, but the material of which they are made was present—albeit in rather different form!—at that event.

same conclusion on the basis of *his* observations of the motions of the other galaxies relative to him.

Using the value $H_0 = 55$km/sec/Mpc we find that the "age of the Universe", the time since the Big Bang, is about 18 billion years (fig. 46). This value would be correct only if there were no matter in the Universe, for the effect of the mutual gravitational attractions of the galaxies would be to slow down their rate of recession. It is reasonable to suppose, therefore, that the galaxies were separating more rapidly in the past than at present, and to deduce that the age

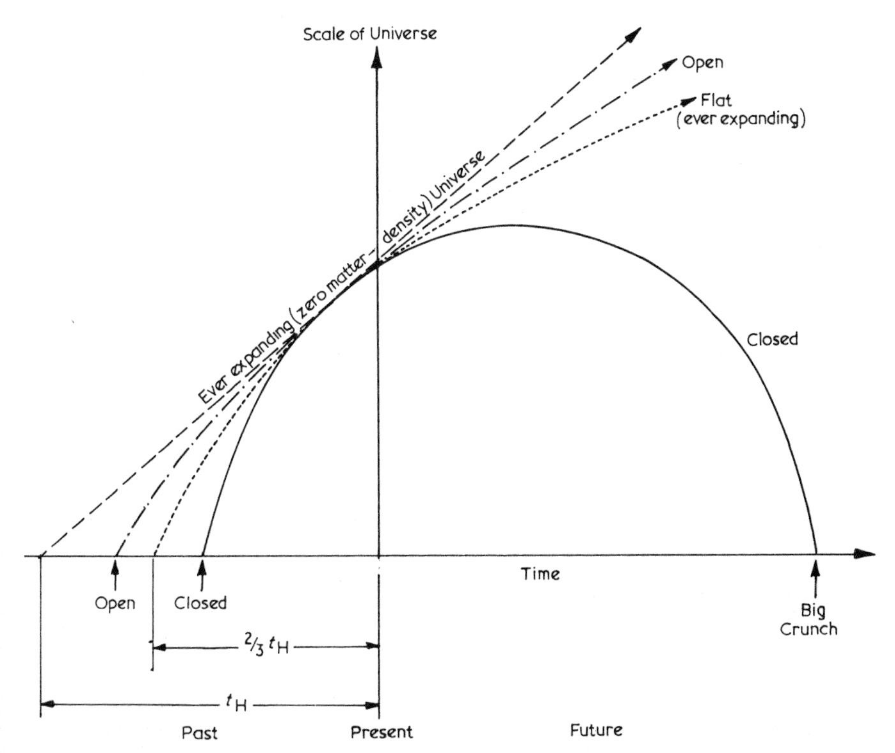

Fig. 46 *The age of the Universe—Friedmann "Big-Bang" models.* If the Universe is expanding at a constant velocity, its "radius" must have been zero at a finite time in the past; the interval between that time and the present is known as the Hubble time (t_H) and, assuming the value of Hubble's constant referred to in the text, is about 18 billion years. However, matter is present in the Universe: thus the expansion must be slowing down, and so the Hubble Age is likely to be an upper limit. If the Universe has precisely the critical density, it will *just* expand without limit (i.e., the velocity of expansion will forever decrease, but never quite to zero), and the present age will be $\frac{2}{3}t_H$ (about 12 billion years). A closed Universe will be younger than this; while an open Universe will have an age between t_H and $\frac{2}{3}t_H$.

of the universe should be less than the value arrived at above. For example, the "flat" Friedmann model of the Universe, which is only just capable of expanding to infinity, would have an age of about 12 billion years; while if the Universe is closed, fated eventually to fall together again, the age would be less than this value. For an ever-expanding, open, Universe the age would lie between 12 and 18 billion years.

Hubble's original measurements yielded a value of H_0 of about 550km/sec/Mpc, about ten times greater than the presently favoured value. Distances based on that value would be ten times less than presently estimated distances (a distance can be assigned to a galaxy, assuming Hubble's law holds good at large ranges, on the basis of its recessional velocity as calculated from its red-shift), and such a value would lead to an age for the Universe of only about 2 billion years, a value far less than the estimated age of the Earth (4.6 billion years) or the oldest observed star, which appears to be at least 10 billion years old.

It is difficult to conceive of a Universe younger than its constituent bodies, and the resolution of this apparent discrepancy was one of the *raisons d'être* for a radical alternative to Big-Bang cosmology put forward in 1948 by H. Bondi and T. Gold, and by F. Hoyle: the *Steady-State theory*. This model incorporated the so-called Perfect Cosmological Principle, whereby the Universe looked the same everywhere and *at all times*. Such a Universe would be infinite in space and time—it would have no beginning and no end. In the Steady-State model, galaxies were receding from each others with velocities which increased in proportion to their increasing separations, a feature which fitted the observational evidence and which avoided Olbers' Paradox. The penalty to be paid was the violation of the conservation of mass-energy for, in order to maintain the same overall average density of galaxies in any region of space at all times, matter had to be continuously created in order to generate new galaxies to take the places of those which were separating from each other. The concept of the continuous creation of matter was attacked as philosophically unsatisfactory, although it could be said that it was no more unsatisfactory than the sudden appearance of all the matter in the Big Bang. Direct observational tests would not be possible, since the required creation rate amounted only to about one hydrogen atom per cubic metre per billion years.

Throughout the nineteen-fifties and early -sixties the debate ranged hotly between protagonists of the opposing theories, but by the mid-'sixties evidence had begun to accumulate which was to negate the Steady-State theory.

One way of attempting to distinguish between the theories was to carry out counts of the numbers of galaxies or radio sources visible at increasing distances (or at increasing red-shifts). By looking to greater and greater distances, astronomers are looking further and further back in time: according to the Steady-State theory the mean separation of the galaxies should be the same in the distant past as they are now, while according to the Big-Bang theory galaxies would have been closer together billions of years ago. Such differences should show up in the way in which the number of sources in the sky increase as we count fainter and fainter (more and more distant) objects. To make the distinction, it was necessary to look to distances beyond the range at which ordinary galaxies may be detected, and so the source counts concentrated on radio galaxies. The results obtained in these surveys indicated that the number of faint sources increased more rapidly than the predictions of the Steady-State theory would allow.

For a time the interpretation of these results was questioned, but then came an accidental discovery which—more than any other piece of evidence—led to the downfall of the Steady-State theory. While testing a sensitive communications antenna during 1964 A. A. Penzias and R. Wilson of Bell Laboratories discovered a weak background level of microwave radiation which, despite their best efforts, they could not eliminate. In 1965 they concluded that this *microwave background radiation* was coming from space, and that it was highly isotropic. Curiously enough, at about that time R. Dicke, P. G. Roll and D. T. Wilkinson at Princeton had begun to construct an antenna system with the specific intention of looking for radiation of that kind for, according to calculations carried out by various theorists since the early work of George Gamow (1904–1968) in 1948, the Universe should contain a weak background of microwave radiation if indeed it began in a hot dense state.

After the initial event, the Universe would have consisted of a very hot dense "soup" of matter and radiation which would cool down as it expanded. After a time (about 700,000 years, according to current estimates) the temperature would have dropped to about

4,000 K, sufficiently cool for protons and electrons to come together to form electrically neutral hydrogen atoms. When this happened the expanding soup, which hitherto had been opaque to electromagnetic radiation, would have become transparent (neutral atoms interact with radiation far less effectively than do charged particles) and radiation would have gone its own way, spreading throughout the expanding Universe (fig. 47). When released, the radiation would have had the same kind of character as black-body radiation released from a hot body at a temperature between 3,000 and 4,000 K; i.e., the radiation would be similar in nature to that emitted by a reddish, moderately cool star.

The radiation released from the Big Bang would pervade all space, and therefore would be expected to be highly isotropic. The observed black-body spectrum and temperature of the microwave radiation agrees very well with the predictions of the Big-Bang theory; its existence was not predicted by and cannot be explained by the Steady-State theory. In the most dramatic way this chance discovery showed the Steady-State theory to be untenable.

Because of the expansion of the Universe, the primordial radiation has been severely red-shifted. The radiation which we are now receiving was released some 700,000 years after the Big Bang, and to look back in time to the era when it was emitted we have to look into the depths of space to a distance corresponding to a red-shift of about 1,000. This enormous red-shift has moved the peak wavelength of the primordial radiation from the red end of the visible spectrum into the microwave region, the peak of the presently observed background radiation occurring at a wavelength of about a millimetre. The effective temperature of this red-shifted radiation is now about 3 K, compared to its original value of around 3,000 K when it emerged from the expanding Big Bang fireball.

Penzias and Wilson shared part of the 1978 Nobel Prize for Physics in recognition of the significance of their work.

Another key issue favoured the Big Bang at the expense of the Steady State, and that was the "helium problem". By and large, wherever astronomers look in the Universe they find that the relative abundances of the two lightest elements, hydrogen and helium, are the same, probably about 25–30% of the matter in the Universe being helium and most of the rest being hydrogen, the simplest element of all. The problem was two-fold: where had all the helium

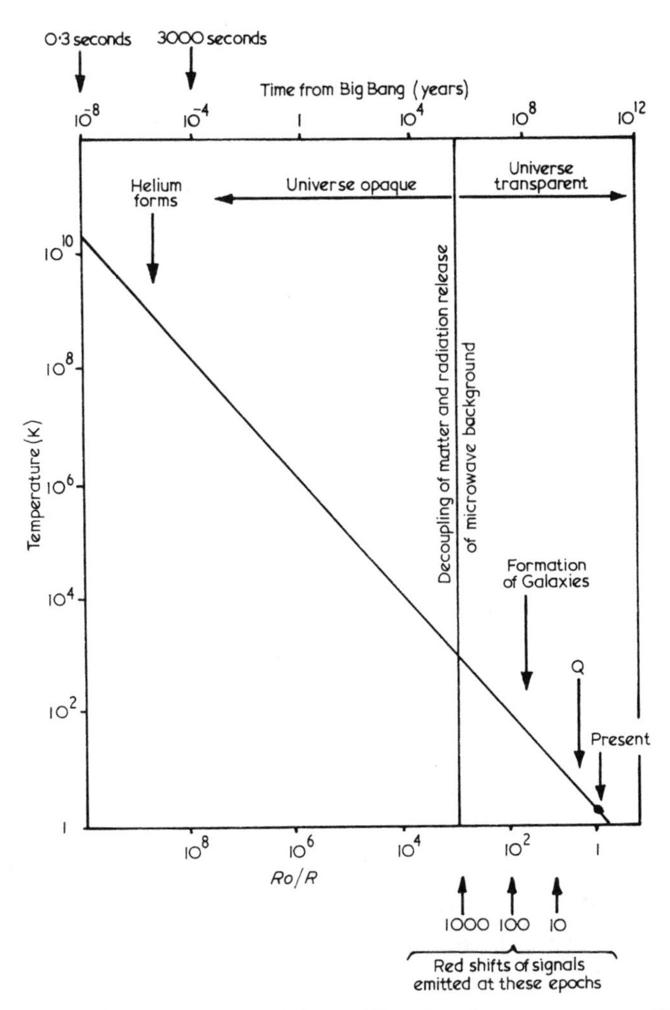

Fig. 47 *The thermal history of the Universe.* The changing temperature of the Universe (or of its radiation content, at least) plotted against time from the Big Bang (the straight line is *an approximation* only). The lower scale shows the value of the present "radius", or scale factor, of the Universe (R_0) compared to the value R at earlier times. When the Universe was about 1 second old, the temperature of matter and radiation was about 10 billion K (i.e., 10^{10} K); today the temperature of the background radiation is just under 3 K. At some 700,000 years after the Big Bang, the Universe became transparent and the background radiation was released (matter and radiation "decoupled"). At that time the Universe had about one thousandth of its present "radius"; and the signals received from that era have been red-shifted by a factor of about 1000. The red-shift of the most distant quasar (Q) is about 3.6; we do not at present have direct observations relating to the period corresponding to red-shifts between 1000 and 3.6—that is, between about 700,000 and 2 billion years after the Big Bang.

come from, and why was it distributed so uniformly? Helium is manufactured inside stars in the nuclear reactions which keep them shining, but only a small proportion of stars become supernovae, scattering their nuclear products into the interstellar clouds from which new generations of stars are born. So the observed quantities of helium cannot have been produced inside stars, as would be necessary for the Steady-State theory, resting as it does upon the proposition that the atoms created in space are of hydrogen. According to the standard model of the Big-Bang, conditions in the first few minutes of the expansion were ideal for the production of helium in the relative proportions in which we find it today. The fact that helium would have formed at the same time throughout the fireball would have ensured its uniformly distribution today.

In the light of the available evidence—source counts, microwave background radiation and the helium problem—it is hardly surprising that by the early nineteen-seventies the Steady-State theory had been abandoned, and the hot Big-Bang theory had come to assume an almost dogmatic status.

The history of the Universe, according to the "standard" Big-Bang model

At time zero the Universe emerged from a singularity. During the first millionth of a second, when the temperature was well in excess of 10^{12} K and the density was beyond our comprehension, exotic interactions beyond the ken of present-day physics must have been taking place at an inconceivable rate. We can only speculate at present on what went on in those first instants; perhaps, for example, the four forces of nature were as one, in the beginning. We have every reason to suppose, however, that by the end of the first millionth of a second there existed a primeval soup of hot energetic particles—particles of radiation (photons) and particles of matter. The whole interacting mass was in a state which we call thermal equilibrium.

Einstein's famous equation $E=mc^2$ tells us that matter can be converted into energy and energy into matter. Thus in those early instants all kinds of particles must have been forming and annihilating. Any material particle has a certain amount of mass, and so a certain minimum "'threshold energy" is required for its formation; so long as the energy density of photons was high enough, any kind of particle could be created. We know, too, that when gamma rays (energetic photons) create particles they create them in pairs comprising a particle and its antiparticle—e.g., and electron and a positron. In the dense conditions prevailing in the early Universe particles and antiparticles would immediately collide again, forming gamma radiation. This interplay between particles and radiation would continue for as long as the energy density was high enough for the photons to have energies in excess of the particle thresholds.

By the time the Universe was one hundredth of a second old, the temperature had dropped to about 10^{11} K, well below the threshold at which protons and neutrons could be created, but some of these particles must have survived the mutual annihilation process—otherwise there would be no matter in the Universe today! After 1 second, the temperature had dropped to about 10^{10} K and neutrinos had ceased to interact significantly with matter; in effect, the Universe became transparent to neutrinos. Electrons and positrons were still annihilating and reforming, but after 10 seconds or so the energy level dropped below their threshold, too, and the great majority of positrons and electrons wiped each other out in an orgy of destruction, leaving behind only a small residue of electrons, eventually to link up with protons and neutrons to give rise to the material content of the Universe as we know it today.

There must have been a slight imbalance between matter particles (protons, neutrons, electrons, etc.) and antimatter particles (antiprotons, antineutrons, positrons, etc.), since otherwise all "material" particles would have been destroyed. In our part of the Universe, matter exists in overwhelmingly greater quantities than antimatter, which is found only in the form of individual particles. It is possible, of course, that in those early instants the Universe had regions where ordinary matter dominated and regions where antimatter dominated—in which case there may be antimatter stars and galaxies; at a distance these would be indistinguishable from stars and galaxies made from ordinary matter. However, there is no generally accepted evidence to support such a contention, and it is more reasonable to assume that there was a small but significant overall imbalance. Indeed, there are theories under development at present (Grand Unified Theories—see Chapter 12) out of which this imbalance appears to come naturally.

By the time the Universe was about 3 minutes old, the temperature had dropped to about 10^9 K and conditions were right for the formation of helium, practically all the available neutrons being mopped up by this process. About 1 minute later, the material content of the Universe consisted almost entirely of nuclei of hydrogen and helium in much the same proportions as we observe today. Thereafter the expansion of the primeval soup continued with very little change until, after about 700,000 years, electrons and protons were able to combine to form neutral hydrogen atoms, the Universe became transparent to radiation, and what we now observe as the microwave background radiation was released.

When matter became transparent to radiation, gravity came into its own, dominating the interactions between the largely neutral masses of material making up the bulk of matter in the Universe. Gravity caused galaxies, clusters, stars and planets to form out of the primeval material released from the rapidly thinning and cooling fireball, and gravity will determine the evolution and ultimate fate of the Universe itself. However, there are a great many unanswered questions relating to the era following the decoupling of matter and radiation, and to the question of galaxy and star formation. Did galaxies form before the first generations of stars, or *vice versa*? How did matter manage to form into discrete blobs—stars, galaxies clusters and superclusters—while the Universe as a whole was rushing apart?

There are two basic schools of thought on the way in which galaxies were formed. One point of view is that at any instant there would exist, at random, regions of greater-than-average density within the expanding mixture of matter and radiation: due to the effects of gravity on these, there would form at first very large blobs, within which fragmentation would occur, leading to smaller entities which became the clusters and individual galaxies which we see today, then, within the galaxy-sized blobs, star formation took place, again as a result of gravity acting on regions of locally enhanced density (see Chapter 6). The other viewpoint involves a different order of events: at first a great many small galaxies formed from fluctuations in the density of matter in the expanding soup, and as time went by these gathered together into clusters, superclusters and, perhaps, even larger hierarchical groupings.

Central to this debate is the question of just how turbulent or

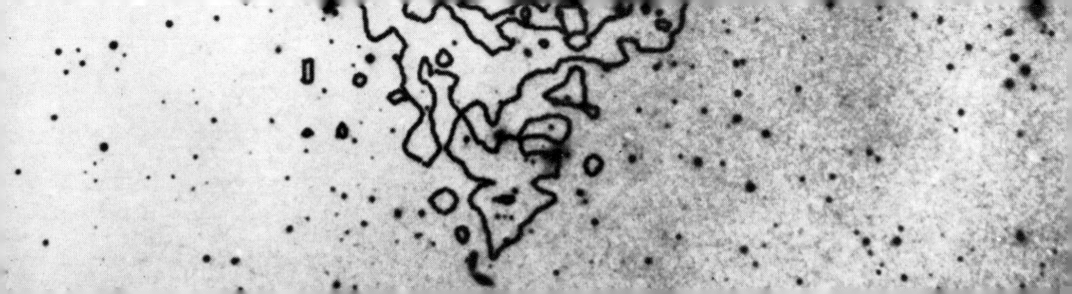

Stephan's quartet. A compact group of four galaxies, once considered to be a 'quintet'; but recent observations indicate that one of the five is a foreground galaxy. In common with other groups and clusters it appears as if there is insufficient mass contained in the visible galaxies to hold the group together. The gravitational attraction of invisible 'missing' matter appears to be required to prevent the group's dispersing. (Lick Observatory Photograph.)

smooth the Big Bang was. The Universe today seems remarkably smooth on the large scale: although there are some irregularities, the distribution of distant galaxies and clusters is highly uniform, over the sky as a whole, while the background radiation is isotropic to better than one part in 3,000. On the face of it, this would seem to imply that the Big Bang itself was a remarkably smooth and well ordered state of expansion; but, in that case, how did the density fluctuations necessary for galaxy formation come about? Our difficulty at present is that we do not have observational data relating to the crucial period of galaxy formation. According to the conventional view, the microwave background gives us information concerning the time when the Universe was some 700,000 years old, corresponding to a red-shift of about 1,000; the most distant quasar so far observed has a red-shift of 3.6, which implies that the light which we are receiving from it was emitted when the Universe was just under 2 billion years old. Between these two times, a great deal must have happened—including galaxy formation. Nevertheless, the available evidence at the moment tends to favour the second hypothesis mentioned above, that galaxies formed first and then congregated into clusters and superclusters.

Because of the success of the Big-Bang model, it tends to be taken for granted that the microwave background radiation really does represent the radiation released from the expanding fireball when the Universe became transparent. Perhaps this is too simple a view. In 1978, in an attempt to account for the fact that the observed ratio of photons to baryons ("heavy" particles, such as protons and neutrons) is about 10^8:1, M. Rees suggested that the observed background radiation could be explained if there had been a burst of massive star formation just after the "decoupling" and before the Universe was a million years old. Such stars would live for at most about 10 million years and many would explode as supernovae and release heavy elements, some of which would condense to form solid particles, making up clouds of dust. The dust heated by this early pregalactic population of stars would have emitted infrared radiation which, allowing for the red-shift due to the expansion of the Universe, would appear today as a microwave background.

This is an unconventional view, but it is interesting to note that in 1979 D. P. Woody and P. L. Richards of the University of

California published measurements which appeared to show deviations from the ideal black-body curve in the microwave background; in effect the curve seemed to be "sharper" than it ought to be. Later that year it was pointed out by M. Rowan-Robinson, J. Negroponte and J. Silk (Queen Mary College, London) that Woody and Richards' "hump" in the microwave background could be explained by emission from clouds of dust formed after a burst of star formation along the lines suggested by Rees. It is too early to say whether this novel idea will stand up to further investigation; but, if it is correct, it implies that the great majority of all the mass in the Universe is contained in the dead remnants of this earliest pregalactic generation of stars, and may now lie in massive dark haloes around the luminous galaxies which we see today.

The future of the Universe

Leaving aside the vexed question of galaxy formation, let us look at what current theory and observation has to say about the future development and probable fate of the Universe.

There is no doubt that gravity will control the future course of events. Is there sufficient matter* in the Universe for the net gravitational attraction eventually to halt the expansion and cause the galaxies to fall back together again, so that the Universe ends in a "Big Crunch"? Or, alternatively, will the Universe continue to expand without limit?

We can think of the expansion of the Universe in terms of the familiar concept of escape velocity. In Newtonian gravitational theory the net attraction on a particle placed inside a hollow spherical shell of matter is zero—the attractions from different parts of the shell cancel each other. The same holds true in General Relativity. Consequently, if we select for examination a typical spherical region of the Universe, the rest can be regarded as a thick hollow shell outside the selected region—because, in accordance with the Cosmological Principle, the Universe looks the same in every direction and matter must be evenly distributed in every direction. We can

* Strictly speaking, we should talk of the total mass-energy of the Universe, for energy, too, contributes an effective quantity of mass to the universe. However, present evidence suggests that matter is by far the dominant factor in determining the open or closed nature of the Universe.

consider a galaxy at the edge of the selected region (fig. 48) to be acted upon only by the matter contained *inside* that spherical region. If that matter is evenly distributed, the galaxy will experience an attraction towards the centre of the sphere, as if all the mass contained in the sphere were concentrated there. In its motion relative to the centre, the galaxy will behave like a projectile thrown outwards from that point. If the galaxy has a velocity sufficiently great—i.e., if it exceeds the "escape velocity" of the spherical region—it will continue to move away forever (and the Universe will be open), but if it has not its velocity must eventually drop to zero, after which it will be dragged back to the centre (and the Universe will be closed).

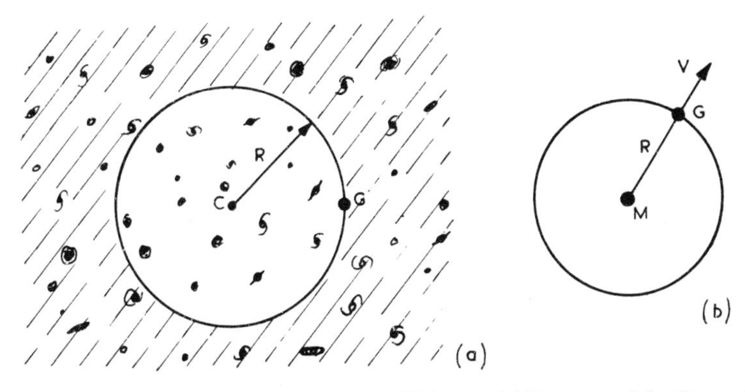

Fig. 48 *Mean density and the expansion of the Universe.* (*a*) Because of the Cosmological Principle, we can select any representative spherical region of the Universe and ignore the gravitational influence on galaxy G of all the material which lies outside that region (see text). Consequently (*b*), we can regard the motion of G subject to the attraction of the matter within the sphere as if that matter were all concentrated in one central mass M. Knowing what mass is required to prevent the "escape" of galaxy G, we can calculate what mean density of matter is required to prevent this region, and the Universe as a whole, from expanding without limit.

Knowing the rate at which the galaxies are separating—this is given by Hubble's constant—we can work out how much mass has to be contained within a given volume of space in order to halt the expansion; i.e., we can establish the mean density of matter necessary to imply a closed Universe. If the mean density exceeds a certain value, known as the *critical density*, then the Universe will eventually cease to expand, and gravity will have won the battle: the ultimate collapse of the Universe will be assured.

Taking H_0 to be 55km/sec/Mpc, the value of the critical density turns out to be about 5×10^{-27} kg per cubic metre, or, on average, about 3 hydrogen atoms per cubic metre—a very low value, but then the Universe is a very large place and its matter is very thinly spread out. The determination of the mean density is one of the crucial tasks in astronomy today.

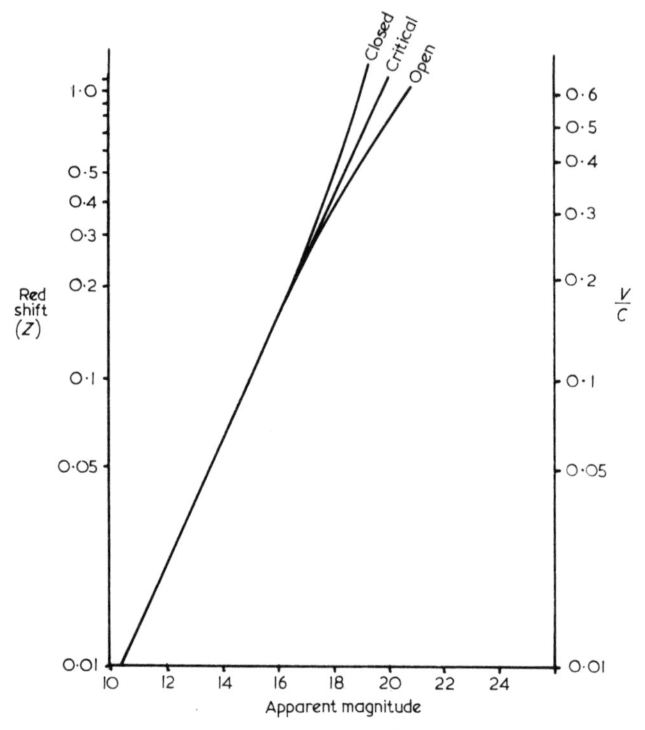

Fig. 49 *Hubble's law for various model universes.* The various possible models for the expanding Universe give rise to differing Hubble laws at large red-shifts. Here red-shift (and, on the right, velocity of recession expressed as a fraction of the speed of light) are plotted against apparent magnitude (a measure of observed brightness which, in turn, is a measure of distance) for galaxies. If the Universe is closed, galaxies will have been moving faster in the past than if the Universe is open. However, it is only at large red-shifts that these differences begin to become apparent, and results to date are good enough only to rule out a *very* rapidly decelerating Universe—which would, obviously, be closed.

Another approach to the problem of determining whether the Universe is open or closed is to try to measure directly the deceleration of the expansion; i.e., to measure what is known as the *deceleration parameter* (denoted by q_0). By looking to very great distances, we are looking back in time to epochs when—if the Big-Bang theory is correct—the Universe must have been expanding more rapidly than it now is. In principle, by measuring the distances and red-shifts of galaxies at the greatest possible ranges, it should be possible to find a deviation from Hubble's law for the most remote galaxies (fig. 49). In practice it has proved impossible, so far at least, to obtain reliable and consistent results from this approach. There are many difficulties, including the problem of reliable distance estimation, and allowance for unknown evolutionary effects: for example, were the galaxies more luminous in the past than they are now and, if so, by how much? To have any hope of distinguishing between open and closed models it is essential to look to red-shifts greater than 0.5, and this means looking beyond the range at which ordinary galaxies can be seen at present (the Space Telescope, scheduled to be launched during the nineteen-eighties, may alter this situation). Quasars are the obvious sources to study, but there is so much uncertainty about their nature, distance and evolution that any results obtained remain suspect. To date, all likely models—open or closed—have the support of *some* of the measurements!

Attempts have been made to determine the age of the Universe by diverse means and to compare the results with the Hubble time—the age which the Universe would have if there were no deceleration (about 18 billion years, assuming H_0 to be 55km/sec/Mpc). Estimates of the ages of the oldest stars in globular clusters, based on their chemical compositions and current theories of stellar evolution, yield values ranging between 8 and 18 billion years, while arguments based on radioactive decay place a lower limit of about 6 billion years. Taking the evidence available to them in 1978, D. Kazanas and D. N. Schramm of the University of Chicago and K. Hainebach of the University of California concluded that the age most consistent with the available data lay between 13.5 and 15.5 billion years, a value consistent with an open, ever-expanding Universe.

On the other hand, in 1977, D. Lynden-Bell of Cambridge obtained a value for H_0 of about 110km/sec/Mpc, as a result of his analysis of a model which he developed to account for the apparent superluminal separation velocities of radio components in certain quasars. Such a value, if correct, would imply that the Hubble age of the Universe is only 9 billion years, a value which verges on being inconsistent with the ages of the oldest known stars.

In fact, if deceleration is taken into account, very real problems arise in trying to fit the measured ages of stars into a simple Big-Bang model. Results published in 1979 by D. Hanes of Cambridge indicate a Hubble age of about 13 billion years, while in the same year M. Aaronson of Steward Observatory, J. Huchra of Harvard University and J. Mould of Kitt Peak National Observatory published results based on infrared measurements of the luminosities of galaxies which indicated an age of about 10 billion years ($H_0 = 100$km/sec/Mpc).

More recently, in 1980, results of an analysis of the radioactive element rhenium in meteorite samples were published by J. M. Luck, J. L. Birk and C. J. Allende of the University of Paris. Rhenium has a very long half-life (half of any sample will decay to form osmium in a period of 60 billion years), and by comparing the amounts of rhenium and osmium in meteorites—assuming the rhenium to have been formed in supernovae at an early epoch—the authors concluded that the age of the Universe lies between 13 and 22 billion years.

At present, although it is probably true to say that the majority of astronomers favour the value of H_0 which yields a Hubble age of some 18 billion years, there is wide disagreement, and it is therefore not possible to compare to Hubble age with the ages of the constituent parts of the Universe in order to assess the rate of deceleration.

The mean density of the Universe

For the moment, the most hopeful approach to tackling the problem of the future evolution of the Universe seems to lie in assessing the mean density of matter. How can this be done?

The obvious approach is to measure the masses of all the galaxies within a given volume of space; dividing this by the volume would yield a value for the mean density.

The rotational motion of stars and gas in a galaxy depends upon the gravitational forces to which they are subjected so that—taking a very simple-minded view—the speeds at which stars in the outer parts of a galaxy travel depend upon the mass of that galaxy. By analysing these motions an estimate for the mass may be obtained. The total luminosity and spectral characteristics of a galaxy also can be used to assess its mass, since we can estimate the number of stars necessary to produce that luminosity; general rules of thumb linking mass to luminosity (M/L ratios) can be established for different types of galaxies. Armed with this information, it is possible to count the number of galaxies in a given volume of space and assess their masses, so leading to a value for mean density.

As with most cosmological investigations, it is difficult to apply these ideas in practice. Uncertainties in the individual masses of stars, and in the distribution of galaxies due to clustering and superclustering, the problems of distance measurement, and the possibility of our failing to see large numbers of faint galaxies—all of these conspire to add a considerable measure of uncertainty to our estimates. Nevertheless, without exception, all available estimates indicate that there is not nearly enough matter contained in the visible galaxies to halt the expansion of the Universe. Current values suggest that the visible galaxies contain 1–4% (although uncertainties could push this up to a maximum of 10%) of the matter necessary to close the Universe.

If all matter were contained in visible galaxies alone, then the Universe would be fated to expand forever. An obvious question to ask is: "*Is* all the matter in the Universe contained in the form of luminous galaxies?" There is compelling evidence available which allows us to give an unequivocal "No" in answer to that question: evidence which is mounting almost daily points to the fact that the visible material seen in galaxies constitutes merely a fraction of the total mass of the Universe.

One of the first clues to the existence of a great deal of invisible matter comes from analyses of clusters of galaxies. One way of measuring the mass of a cluster is to add up the masses of the visible galaxies; with the Coma cluster, for example, the mass turns out to be about 30,000 billion solar masses. Another approach depends upon measuring the motions of the member galaxies within the cluster, and leads to what is known as the cluster's *virial mass.** In

*The *virial theorem* relates the kinetic energies of cluster members to the total gravitational energy of the cluster. If the speeds of the galaxies are too high then they will escape, and the cluster will disperse. By measuring these speeds we can determine the total mass necessary to hold the cluster together.

the case of the Coma cluster, the virial mass turns out to be about a hundred times greater than that obtained by counting the masses of the visible galaxies!

Although this is an extreme example, the fact remains that when clusters and groups of galaxies are investigated in this way it is invariably found that more mass is needed to hold them together than is revealed by mass estimates of the visible galaxies.

What forms might this missing mass take? One possibility is the existence of clouds of gas in the space between the galaxies. Evidence for the existence of intergalactic gas comes in a variety of forms. There are some radio galaxies, such as NGC 1265 in the Perseus cluster, whose shapes seen on radio contour maps show a distinct "head-tail" structure, as if their radio-emitting clouds were being swept behind the main body of the galaxy by a "wind" created by the galaxy's motion through a tenuous medium of intergalactic gas. If this gas were very hot and ionized (made up of charged particles), it would not emit the characteristic 21cm radiation which betrays the presence of hydrogen clouds in our own galaxy, and so we would not be able to detect it directly.

The most convincing evidence for the presence of intergalactic gas in clusters comes from observations of X-ray emission from clusters, this emission coming from the general regions of the clusters rather than from the individual member galaxies. Gas at temperatures of about 10 to 20 million K would emit X-rays of the appropriate kind. No less than 20 of the original sources in the Uhuru satellite's pioneering X-ray survey turned out to be clusters of galaxies, and later generations of satellites have extended and improved our knowledge of sources of this kind. The Coma cluster source, for example, could be accounted for by a cloud of hot gas having a mass of about 5×10^{14} solar masses; however, even this is only about one tenth of the total mass required to hold the cluster together.

The intergalactic matter in clusters so far detected by X-ray observations does not appear to provide anything like enough mass to hold clusters of galaxies together, and certainly does not indicate the presence of sufficient mass to close the Universe. Recent observations of galaxies have hinted at the possibility that galaxies possess massive haloes of material, possibly in the form of "dead" stars which may have formed very early in the history of the Universe (see page 216). It has been proposed that if, as Rees and others have suggested, a large proportion of the material in the Universe was involved in a burst of star formation when the Universe was very young, the assembling of these earliest stars into hierarchical structures may have led to the formation of galaxies, the galaxies which we see today being merely the luminous remnants formed out of gas which accumulated at the centres of massive systems of dead and dying stars. If as much as 90% of the mass of a typical galaxy is contained in its invisible halo, then we have found the mass required in order to hold clusters of galaxies together. However, even should this prove to be the case, the haloes probably do not contain quite sufficient mass to close the Universe as a whole.

It has been argued that there may be a large-scale tenuous distribution of matter throughout the Universe, even between the clusters. The rate at which the pendulum of opinion on this issue has swung to and fro is a measure of the rate of progress of observation and theory. Measurements made in 1978 by the X-ray satellite HEAO-1 indicated a diffuse universal background of X-rays which could be accounted for by the presence of a hot, all pervading plasma (ionized gas), possibly containing enough material to close the Universe. Early the following year, the Einstein Observatory (HEAO-2) showed that a substantial fraction of the X-ray background could be due to individual sources, such as quasars—more than 100 of which were shown to be X-ray emitters. In the light of this evidence, it seemed *most unlikely* that there was sufficient hot gas to halt the expansion of the Universe.

Are there any other possibilities worthy of our consideration? Large numbers of neutrinos or of gravitational waves produced in the Big Bang could carry substantial amounts of energy which would contribute to the curvature of the Universe. The detection of either of these lies beyond our abilities at present—it is hard enough to collect neutrinos from the Sun!—but the general opinion is that these factors are unlikely to exert a significant effect although recent suggestions that neutrinos may have finite rest-masses could dramatically alter the situation.

Black holes would seem to be the ideal repositories of almost unlimited quantities of hidden matter, but they are so hard to detect that it is difficult to set limits on their possible numbers. Nevertheless, certain upper limits *can* be deduced. For example, the amount of gamma-ray background radiation, as we have already seen, sets limits on the possible numbers of primordial black holes: unless mini black holes of less than the Hawking mass (the mass of presently exploding holes) were for some reason prevented from forming, the limit on primordial black holes amounts only to one hundred millionth of the critical mass. At the other end of the scale, significant numbers of hypermassive black holes would appear to be ruled out, as any hole of mass greater than about 10^{15} solar masses would produce gross tidal distortions in galaxies and clusters, as well as appreciable gravitational-lens effects. To date, although about 1,000 quasars are known, there is only one plausible case of image doubling due to the gravitational lens effect, and the lens in that case appears to be a massive galaxy rather than a black hole *per se*.

According to B. Carr of Cambridge who in 1978 assessed the various limits which could be placed on black holes at that time, the lack of image-doubling effects sets a limit on the number of black holes with masses in excess of 10^{12} solar masses; this limit allows them to contribute less than 2% of the critical density. The contribution made to the overall mass by black holes or by other massive bodies depends on whether they are clustered in galaxies or spread throughout all space. In Carr's opinion, black holes could provide enough mass to close the Universe only if they were *unclustered*, which implies that they were formed before rather than within the galaxies.

As we have seen, there may have been a rapid burst of star formation shortly after the Universe became transparent, some 700,000 years after the Big Bang. At this stage it is possible that masses of around a million solar masses could have collapsed directly to form black holes, or could have fragmented into numbers of massive stars which then evolved rapidly to become black holes. Such holes would probably lie in the mass range from about a hundred to a million solar masses and could comprise a substantial fraction of massive galactic haloes. In this context, it is interesting to note that the work by Rowan-Robinson and his colleagues alluded to earlier (page 216) suggests that, if dust from these earliest generations of stars is responsible for the "hump" on the microwave background, the mass of (now dead) stars involved would have had to have been comparable with the critical density.

Another fascinating observation indicative of the presence of hidden mass in our general neighbourhood resulted from some measurements carried out in the late nineteen-seventies from a high-flying U-2 aircraft by G. F. Smoot, M. V. Gorenstein and R. A. Muller of the University of California. The results showed that our Galaxy, together with the rest of the Local Group, is moving relative to the microwave background radiation; the movement is betrayed by a slight blue-shift in the background in the direction in which we are heading, and a slight red-shift in the background "behind" us. The Local Group as a whole is moving towards a point in the direction of the Virgo cluster of galaxies at a speed of about 600km per second. One interpretation is that the Virgo supercluster contains a great deal more matter than we previously thought, while an alternative is that the attracting mass is even larger, and lies beyond the Virgo supercluster. Whatever the truth of the matter, these observations seem to demonstrate the existence of large-scale inhomogeneities in the structure of the Universe, despite its general appearance of uniformity and isotropy. It remains to be seen whether the standard Big-Bang model can accommodate irregularities of this kind.

There is, then, clear evidence for the existence of a great deal more matter than we can see, but precisely how much remains a matter of great debate. If all the preceding techniques of attempting to measure the present matter density are beset with errors of observation and interpretation, is there some other way of tackling the mean-density problem?

It may be that the present abundance of deuterium ("heavy" hydrogen, containing not just a proton but also a neutron in its nucleus) provides a vital clue. According to the standard model, deuterium would have formed when the Universe was about 3 minutes old, its formation being an essential prerequisite to the formation of large amounts of helium. The amount of deuterium that survived the Big Bang depends very strongly on the temperature and density prevailing at the time of helium formation, and that in turn is directly related to the *present* density of matter in the Universe.

It is a difficult task to obtain reliable values for the universal deuterium abundance, but current estimates indicate that about 20 parts per million of the interstellar medium is composed of deuterium, and this is consistent with a present mean density of matter of about

4×10^{-28} kg per cubic metre—less than one tenth of the critical density. If the present density of the Universe were ten times higher, less than one thousandth of the observed quantity of deuterium should have survived. Abundances of "lightweight helium" (^{3}He) and lithium are also quite sensitive to the initial conditions, and the measured amounts of these are consistent with their being insufficient mass to close the Universe. These results, although highly suggestive, depend upon the assumption that the existing quantities of deuterium, ^{3}He and lithium were produced in the Big Bang, and there is still a good deal of uncertainty on that score.

In summary, then, we can say that the measured abundances of the lightest elements, the measured masses of clusters of galaxies, the upper limits imposed on most kinds of black holes, and the absence of clear evidence for large amounts of intercluster matter all tend to favour the view that we live in an open, ever-expanding Universe. The deceleration parameter cannot as yet be relied upon to give a clear indication either way, and the case for a closed Universe rests on uncertainties in existing data and on the possibility of as yet undetected further hidden matter, particles or radiation.

The balance of evidence uneasily favours the ever-expanding model, but it must be pointed out that the trend of discoveries over the past few years has usually been to increase the observed mean density. A few years ago, the evidence clearly favoured an open Universe; today we are less confident, and all that we can say for certain is that, if the Universe is indeed closed, then the mean density does not exceed the critical density by much. But, since the value of critical density depends on the square of Hubble's constant, a constant whose value is not certainly known, we should be wary of being dogmatic.

An oscillating Universe?

If we look at a graph of the "radius" of the Universe plotted against time for a closed Universe (fig. 46, page 205) we see that it is symmetrical: the Universe expands rapidly at first, slows to a halt, and then falls back together again, accelerating all the while until all of the matter and radiation is compressed into a fireball and a final singularity—the Big Crunch. If our Universe turns out, after all, to be closed, will the Big Crunch mark the end of space and time?

The Big Hole:

It has been argued by some that the collapse would lead to some kind of "bounce", resulting in matter and radiation being hurled outwards from a new Big Bang which would herald the start of a new cycle of the Universe. This attractive idea is central to the theory of the *oscillating Universe* (fig. 50), a Universe which expands and contracts in a cyclic way, matter and radiation being reprocessed at the end of each cycle. This "ecologically acceptable" idea has great philosophical attraction, combining as it does the virtues of a finite yet unbounded Universe with the timeless character of the old Steady-

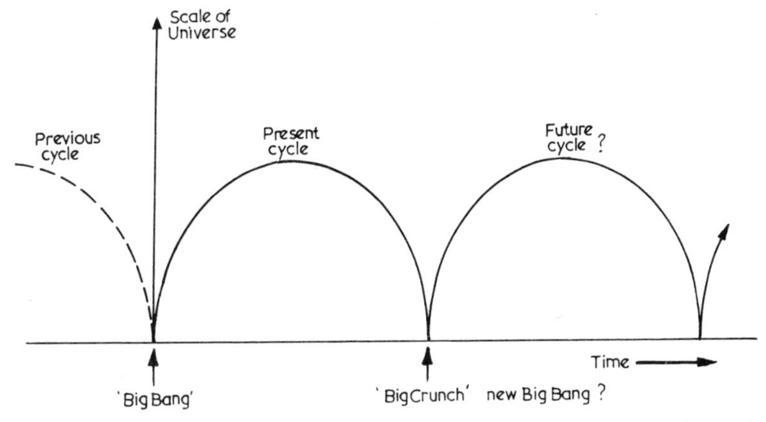

Fig. 50 *The oscillating Universe.* If we live in an oscillating Universe the scale, or "radius", of the Universe should expand and contract in a regular periodic fashion. The collapse at the end of one cycle (the Big Crunch) should herald a new Big Bang and the commencement of a new cycle. If this theory is correct, the expansion and contraction may continue indefinitely in the "future", and there may have been an indefinite number of earlier cycles "prior to" our own.

State hypothesis. Such a Universe would have no unique beginning or ending, just an endless series of cycles.

However, it is no more than an act of faith to suppose that a re-expansion would occur, for there is no known physical law which would allow this to happen. Present-day physics suggests that the Universe would collapse without limit into a singularity just like the one which is assumed to exist inside a black hole. Would the Universe become a black hole, or is it, perhaps, a black hole right now?

If the Universe is closed then it is, in a sense, a "region of space from which nothing can escape". However, it is misleading to attempt to apply the simple concept of a black hole to the Universe as a whole. A black hole is a closed region of space within an external space-time;

i.e., black holes form *within* our Universe. To treat the Universe as a black hole would presuppose the existence of some *external* space-time within which the Universe exists. This may be the case—perhaps there is a limitless multitude of "universes", existing like bubbles in some grander space-time—but we have absolutely no evidence to support it. For the moment we should restrict our application of the term "black hole" to entities within our own Universe.

We are still faced with the problem of the initial singularity, and it is natural—even if it may not be strictly logical—to ask "where" the singularity came from or where the matter and radiation came from. If space and time began with the Big Bang there is no sense in asking what happened "before", as time did not exist prior to that instant. On the other hand, the Big Bang singularity is reminiscent of the concept of a white hole: could it be that our Big Bang was "someone else's" Big Crunch? These are areas of sheer speculation, but we may be confident that the continued investigation of the properties of black holes and their singularities will help us gain some understanding of what went on in that great singularity in our past.

The ultimate fate of an ever-expanding Universe

If, as the balance of evidence suggests, the Universe is fated to expand without limit, what does the future hold? As the expansion continues, matter will become more and more widely dispersed, the galaxies and clusters will become steadily more and more widely separated from each other, and the background radiation will cool towards Absolute Zero. In time all the stars will have passed through their life cycles, ending up as white dwarfs—which eventually cool down to become cold dark black dwarfs—neutron stars or black holes. The era of luminous matter will come to an end, and dark masses, together with elementary particles and radiation, will plunge mindlessly away in an ever-increasing void.

Black holes will still have a rôle to play. Given sufficient time, much of the bulk matter in the Universe will be swallowed by holes. If Hawking's theories are correct, they will be emitting radiation, but for stellar-mass holes it takes a very long time indeed before this has any significant effect. The background temperature will have to drop a long way before such black holes begin even to radiate more than they are absorbing from the cosmic background: this state of affairs will not arise until the Universe is about ten million times older than it is now thought to be. Some 10^{66} years will elapse before stellar-mass holes explode in showers of particles and radiation.

J. D. Barrow of the University of Oxford and F. Tipler of the University of California have painted the following graphic picture of the long-term future of an ever-expanding Universe. Even within a very old neutron star there would be sufficient residual energy to impart from time to time velocities in excess of the escape velocity to particles at its surface; in this way, over a sufficiently long period of time, neutron-stars would be expected to evaporate. Black holes, too, would decay by releasing equal quantities of particles and antiparticles. Barrow and Tipler argue that, if the Universe has only just enough energy to expand forever, the electrical attraction between electron-positron pairs will outweigh both the gravitational attraction and the effect of the general expansion of the Universe, so that within a finite time every electron will find a positron partner with which to indulge in mutual annihilation. That being so, the final state will be not a collection of dark bodies or black holes hurtling away from each other but a tenuous sea of radiation, cooling towards a final uniform temperature.

The second law of thermodynamics predicts that the ultimate state of the Universe is slowly to attain a uniform temperature—as heat passes from hot bodies to cooler ones, all differences become ironed out, and no work is possible thereafter. This concept of the "heat death" of the Universe was first stated by H. L. F. von Helmholtz (1821–1894) as long ago as 1854. It is intriguing and appropriate to find that our modern view of an ever-expanding Universe, coupled with the concept of quantum emission from black holes—an idea which itself followed on from analogies between gravitational physics and thermodynamics—should lead by a more circuitous route to essentially the same conclusion.

We do not know for certain what will be the outcome of the battle between expansion and gravitation. If gravity wins, the Universe will eventually collapse in a Big Crunch, possibly a final event or perhaps the prelude to a new cycle of expansion. If gravity loses the battle, the expansion will go on forever, but gravity will have played a vital rôle in determining the Universe's final state, whether it be a uniform sea of radiation or an expanding collection of dark masses. In the dim and distant future the epoch of stellar activity may seem to be an inconceivably short instant: a mere twinkling of the cosmic eye.

Is this the fate in store for the Universe? It would appear so, but it does seem rather sad and pointless that so marvellous and complex a Universe should become an amorphous dark void. To many, I am sure, the oscillating Universe has much to commend it, offering new hope, if not to living beings, then at least to familiar things like matter and radiation. However, nothing we can do will alter the density or the destiny of the cosmos; we have to accept it as it is. We cannot choose our Universe.

The 'Antennae' – the pair of galaxies NGC 4038 and 4039. Located at a distance of about 100 million light-years these galaxies are so close together in space that their mutual gravitational attraction has greatly distorted their structures, giving rise to the elongated 'feelers', measuring half a million light-years from tip to tip. (Photographed by the UK Schmidt Telescope Unit, Courtesy Photolabs, Royal Observatory, Edinburgh.)

The Virgo Cluster of galaxies. The photograph shows the central regions of this cluster which, at a distance of some 50 million light-years, is the nearest giant cluster. The spiral galaxy NGC 4438 has been greatly distorted as a result of its collision with its companion NGC 4435. In the bottom left-hand corner is the giant elliptical galaxy M87 (see page 160), the nuclear jet of which is not shown in this photograph. (Photographed by the UK Schmidt Telescope Unit, courtesy Photolabs, Royal Observatory, Edinburgh.)

Gravitational wave detector constructed by Professor J. Weber (shown here) of the University of Maryland, USA. The detector consisted of a cylinder of aluminium, weighing some 4 tonnes, which should vibrate with a minute amplitude when acted upon by a gravitational wave. Although Weber's initial results appeared to indicate the detection of gravitational waves, it is now felt that the effects which he was measuring must have been due to some other cause. Although gravitational waves have yet to be positively detected, Weber's pioneering work laid the foundation for a new and potentially exciting branch of astronomy. (Courtesy Professor J. Weber/University of Maryland.)

PART THREE
EPILOGUE

12
The Nature of Gravity

Our discussion of black holes and of the structure, origin and evolution of the Universe has been based upon the best available theory of gravitation, General Relativity. But is it truly the "best theory", or are there different and better theories waiting in the wings, ready to take over its mantle should it falter in the light of experimental tests?

A new theory of gravitation could have profound implications for our conception of the Universe as a whole. Admittedly, General Relativity has withstood all the tests to which it has been subjected in the last six decades, but generally the tests described in Chapter 5 have confirmed the predictions of the theory to only a comparatively modest level of accuracy. This is hardly surprising, in view of the weakness of gravitation in comparison with the other forces of nature; differences between Newtonian theory and General Relativity are bound to be very slight indeed unless one is examining a situation where extremes of gravity are involved—as with a black hole.

Alternative theories exist. In fact, there are several families of theories of different complexions, but most make predictions which differ only in the most marginal degree from those made by General Relativity, at least under the circumstances in which it has been possible to make tests so far. Most of these theories are more complicated than General Relativity, and a number of them require certain of the constants of nature—usually the gravitational constant, G—to change with time; in the absence of compelling evidence in their favour, there is no reason to adopt any one of them in place of a simpler, familiar, well established theory which seems to work so well. Nevertheless, we should be alert to the possibility that some small and apparently innocuous discrepancy between observation and theoretical predictions may lead to the replacement of General Relativity by a more sophisticated theory, in much the same way as the explanation of the orbital motion of Mercury required the overthrow of Newtonian mechanics and gravitation.

Theories of variable G

A central pillar of both Newtonian theory and General Relativity is that the gravitational constant is truly constant for all time. Thus the gravitational interaction between two bodies of given masses would have been the same in the early history of the Universe as it is now.

Is this necessarily so? The first physicist seriously to question the constancy of G was P. A. M. Dirac, of Cambridge University; in 1937 he devised what is known as the "large numbers hypothesis", based upon several striking coincidences in the ratios between key physical quantities.

For example, the ratio of the electrostatic force of repulsion between two electrons compared to the gravitational attraction between them is about 10^{40}:1, a very large number. If we compare what might loosely be called the "radius" of the Universe with the radius of an electron (again, since an electron is not *really* a tiny billiard ball, the term "radius" should be treated with some caution) we find, once more, that the ratio is about 10^{40}:1. (Looking at this latter comparison in another way, we can compare the age of the Universe, which lies between 10^{17} and 10^{18} seconds, with the time taken for light to cross an electron, about 10^{-23} seconds, and we see again that the ratio is about 10^{40}:1.) The first ratio involves certain fundamental atomic constants, while the second involves those same fundamental constants and Hubble's constant, which determines the size and age of the Universe.

Dirac felt that this was no coincidence, and that these ratios showed a link between the values of fundamental constants and the age of the Universe. Comparing the quantities involved in the two ratios indicated that the value of G should be inversely proportional to the age of the Universe, as time went by, gravity would decrease; in other words, as the radius of the Universe increases the second ratio becomes greater than 10^{40}:1, so that if the two ratios genuinely are connected then the first ratio should also increase with time with the result that the gravitational force should be decreasing relative to the electrostatic force.

This was a purely philosophical argument based on numerical coincidences, and was of a type of which Pythagoras and Kepler surely would have approved.

Another philosophical concept which might be interpreted as indicating that gravitational forces should decrease with time is Mach's principle (see Chapter 4, page 59), which states that the inertia of a body is due to the influence of distant masses in remote parts of the Universe. If, as was argued by, for example, D. Sciama (then at Cambridge) in the nineteen-fifties, this interaction was gravitational in nature, then it would be reasonable to expect the interaction to decrease with the increasing size of the Universe. If inertia were to decline, G should do so also, for the equivalence principle requires gravitational and inertial masses to be in exact proportion.

Of the various theories to be developed which incorporate the concept of variable G, one which aroused considerable interest was the Brans-Dicke scalar-tensor theory announced in 1961 by C. Brans and R. Dicke of Princeton University. Adopting the standpoint of Mach's principle, the theory suggested that the local value of G is determined by the structure of the Universe, and it allowed the possibility that the value of G might vary from place to place and from time to time. The equations of the Brans–Dicke theory were similar to those of General Relativity but, in order to accommodate Mach's principle, included an extra variable quantity, the "scalar field", which allowed G to vary.

As we saw in Chapter 5, one of the classical tests of General Relativity is the advance of the perihelion of Mercury. Einstein's theory accounts in a most satisfactory way for the discrepancy of 43 seconds of arc per century between the predictions of Newtonian theory and the observed shift of the orbit. In the calculations, it was assumed that the Sun is a perfect sphere; but any oblateness (flattening of the poles) possessed by the Sun would affect the motion of Mercury. Dicke and H. M. Goldenberg carried out a series of measurements in 1966 which appeared to show a difference between the equatorial and polar radii of the Sun amounting to about 0.04″, a difference which, although tiny, would be sufficient to produce a significant effect. According to the Brans–Dicke theory, the perihelion advance due to relativistic effects would be 39″ per century, the remaining 4″ being caused by the oblateness of the Sun. Dicke argued that his observations favoured the scalar-tensor theory against General Relativity.

Observations made in 1973 by Professor H. Hill and his colleagues have shown the difference in radii to be only about one fifth of Dicke's value, too small to produce the effects predicted by the Brans-Dicke theory. Although there is still room for doubt in the interpretation of the observations, it is generally felt that General Relativity has survived that particular challenge and emerged unscathed. However, scalar-tensor theories are not ruled out in the current state of knowledge.

A quite different kind of theory, starting essentially from Mach's principle and asserting that the influences of distant parts of the Universe are important locally, was proposed in 1964, and in a modified form in 1971, by Sir Fred Hoyle and J. V. Narlikar. Central to the theory was the concept that the mass of a particle is determined by the action of distant particles; in the later formulation the idea was introduced that distant particles could make a positive or a negative contribution to the mass of a local particle. If the Universe consisted of regions, or "aggregates", where the contribution to mass was positive and aggregates where the contribution was negative, then at a boundary where positive and negative contributions cancel out particles would have zero mass.

With the passage of time, the world lines of those particles (and for "particles" we could equally well read "galaxies") would lead away from that boundary, with the result that the net effect of the positive and negative mass contributions would no longer cancel, and *the masses of the particles would increase*. Since the radius of an atom is determined by the masses of its constituent parts, an increase in its mass would lead to a decrease in its size. Hoyle argues that if the fundamental unit of scale, the atom, were shrinking, then the distances between galaxies would *appear* to be increasing even although they might actually be remaining constant. A similar line of argument—that we could not distinguish between on the one hand an expanding Universe with decreasing G and on the other the effects of shrinking atoms—was discussed in the nineteen-forties by E. A. Milne.

The Hoyle–Narlikar theory accounts for the observed red-shifts of the distant galaxies as being a consequence of the changing masses of atoms. The lower the mass of a particular kind of atom the longer the wavelength of the radiation which it would emit. In looking to large distances, we are looking back in time to eras in which atoms were less massive than now, and so the radiation which we are at present receiving from the distant atoms is of longer wavelength than the radiation emitted by the same kinds of atoms locally. In other words, progressively more distant atoms exhibit a progressively greater red-shift.

This radical and unconventional theory suggests that the Universe is vastly greater in extent than we can observe, and is in direct opposition to the conventional view that the Universe began in a singularity a finite time ago. The theory further implies that we cannot receive radiation directly from galaxies which may exist beyond a boundary at which particles have zero mass, because such particles would strongly absorb this radiation. However, Hoyle and Narlikar argue, the particles would reemit radiation, which would appear to us as background radiation! Although novel and fascinating, the Hoyle–Narlikar theory is far removed—for the present at least—from the mainstream of current thinking on the nature of gravitation, and has not received a great deal of support.

Although they represent a minority view, theories which suggest that gravity may vary with time are under active consideration at present. In most theories the rate of change of G is expected to be directly related to the rate of expansion of the Universe which—according to current estimates of Hubble's constant—would amount to between 5 and 10 parts in 10^{11} per year, a very small rate of change.

How might we hope to assess whether or not G is changing? If G were greater in the past, the orbits of the Moon around the Earth and the Earth around the Sun would have been smaller in the past than they are now; in other words, as the strength of gravity diminished orbits would slowly expand. Likewise, the mean angular motion of bodies on their orbits would slow down with time. The computation and measurement of such effects are complicated by the tidal interactions between the Earth and the Moon, which act as a brake on the rotation of the Earth, causing the length of the day to increase and the Moon to recede slowly from the Earth, with a resultant increase in the lunar orbital period. Recent estimates indicate that the increase in orbital period due to this interaction amounts to between 10 and 15 parts per 10^{11} per year. Analysis of ancient eclipse data—changes in the lunar period and the length of the day will affect the times and places from which eclipses were observed—by P. M. Muller of NASA's Jet Propulsion Laboratory and F. R. Stephenson of the University of Newcastle (UK) indicate, according to Muller's calculations, that the average rate of the Moon's recession is 4.4 cm per year: over the 2700-year period investigated, the Moon has receded by about 100 m and the length of the mean solar day has increased by about 0.05 seconds.

In recent years, new techniques have been made available which permit the making of much more precise measurements. In particular, there is the technique of lunar laser-ranging, whereby a very precise distance to a point on the lunar surface can be established by bouncing a laser beam off specially designed reflectors left there for that purpose by the Apollo astronauts. Likewise radar beams can be reflected from the surfaces of the planets, allowing distances to be accurately determined. The great benefit of this technique is that the accuracy increases rapidly with the length of time over which such observations are conducted. Values for the rate of recession of the Moon obtained in this way are around 3.5cm per year.

The difficulty about trying to measure temporal changes in G is that, until recently, our basic standards of time measurement have been based on G itself. The first natural clock used by Man was the rotation of the Earth on its axis; but this does not have a constant rate. A more uniform standard of time measurement is *ephemeris time*, based on the orbital motion of the Earth around the Sun, the year for this purpose being the so-called "tropical year" containing 365.2422 days, each comprising 86,400 ephemeris seconds.

The Earth's orbital period is dependent upon GM, the gravitational constant multiplied by the mass of the Sun. If G were decreasing, the orbital period of the Earth would increase in just the right proportion to prevent the detection of effects due to changing G; if the orbital period of the Earth were to increase, the year, by definition, would still contain the same number of ephemeris seconds, but the "seconds" would be longer. The increase in the orbital period of the Moon would be in the same proportion as the increase in the orbital period of the Earth, and so would be undetectable.

On 1 January 1958 a new standard of time measurement was adopted which is independent of the rotation of the Earth and the motion of the Earth around the Sun—*atomic time*. This timescale is based on the frequency of a particular transition of atoms of the element caesium, that frequency being 9,192,631,770 Hertz (vibrations per second). On this basis time can be measured and regulated to an accuracy of one part in 10^{12}. Atomic clocks subsequently developed can improve on this precision, for limited periods, by a factor of 1,000.

Using atomic time as their standard, T. Van Flandern of the US Naval Observatory and others have analysed twenty years' data on lunar occultations. As the Moon moves around the Earth it passes in front of numerous stars, blotting out their light temporarily; such an event is called an occultation. If the position of the Moon is calculated on the assumption that G is constant then, in principle, it should be possible to predict the time of an occultation with great precision, but if G is decreasing the occultations should occur later than predicted, because of the slowing down of the Moon's angular motion. Van Flandern's results indicate that the Moon's orbital period is increasing, as a result of all causes, by about 22 parts in 10^{11} per year. This appears to be a greater rate than that predicted by our present knowledge of the tidal interaction, and Van Flandern interprets this as indicating a decrease in G of about 3.6 parts in 10^{11} per year, assuming that G is the only "constant" which is changing. If other basic units of mass, length and time are changing, the rate of change of G would be double, about 7.2 parts in 10^{11} per year.

On the face of it, these results appear to be in broad agreement with the rate of expansion of the Universe, but the possible errors are large, and one cannot be sure of estimating accurately all the factors which would affect the motion of the Moon. Apart from the tidal interaction, other factors range from the effects of the solar wind, the relatively close passage of comets, to—as C. Doake of the Scott Polar Institute (Cambridge) has pointed out—the varying amounts of ice contained in the Earth's polar caps.*

Lunar laser-ranging experiments determine the recession of the Moon with respect to atomic time, whereas Muller's results, based on ancient eclipse data, measure this recession with respect to gravitational (epehemeris) time. Muller's results appear to indicate a rate of recession of the Moon about 25% greater than that obtained by the lunar laser-ranging experiments. Is this indicative of a change in G? It would appear so, but on the other hand L. Morrison and C. Ward of the Royal Greenwich Observatory analysed transits of the

*The amount of ice at the polar caps affects, very slightly, the distribution of the Earth's mass which, in turn, affects the gravitational interaction with the Moon.

planet Mercury across the face of the Sun between 1677 and 1973 and obtained a rate of recession of about 3.8cm per year, closely similar to the laser-ranging values; yet these results were based on gravitational time. However, these data do not go back nearly as far as the eclipse data, and the whole question remains open, although the general opinion at the moment is that any change in G must be less than 1 part in 10^{11} per year.

Consequences of variable gravity would include the slow expansion of the Earth and the other planets, and this has been invoked as a possible mechanism to initiate the growth of continents and the process of continental drift. On the other hand, such processes can be accounted for quite adequately without introducing variable gravity. The Earth would have been closer to the Sun and therefore, according to conventional ideas, hotter than at present. As was pointed out by E. Teller, a change in G would affect the interior of the Sun in such a way that the Sun would have been much more luminous in the past: by this reckoning, the oceans would have been boiling, all other factors being equal, as recently as 600 million years ago. This does not tally with geological and biological records.

However, others have argued that the Sun would have been *cooler* in the past if gravity were stronger then, and V. M. Canuto of NASA's Goddard Space Flight Center has pointed out that Newton's and Einstein's equations require GM to be constant. If G were to vary, then to keep GM constant M would have to vary. If this were the case, all the laws used to calculate the effect of changing G on the Sun's luminosity would be invalid. Therefore, he maintains, it is not reasonable to rule out changing G on the basis of conventional arguments which, while calculating the effects of changing G, nevertheless tacitly assume that G is constant!

Together with S. H. Hsieh and P. J. Adams he has put forward a theory which permits the existence of two types of Einstein (relativistic) equations: the first, which would be valid only when gravitational time was used, incorporates GM as a constant, and the second, to be used with the atomic time scale, allows G to vary even if M remains constant; in the latter case GM would *not* be constant. Times measured on the gravitational and atomic scales would be linked by a quantity which itself varies with time; depending on the value of this quantity, atomic clocks would either speed up or slow down relative to gravitational clocks as the age of the Universe increased. In effect, this variable quantity would link atomic phenomena with large-scale gravitational phenomena.

Canuto feels, as have others, including Dirac, that the fact that the electromagnetic force between two electrons is greater than the gravitational force by a factor of about 10^{40} is related to the age of the Universe. If the ratio has always had this value, then why should it be this particular value and no other? Why should the difference be so great? It may be, of course, that the various constants and forces have the values which they do just because, by chance, these were the values which they had when the Universe originated, and we shall never be able to work out why. This seems unsatisfactory. Perhaps at the initial instant the two forces had the same strength, and have since attained their present values as a result of the expansion, or ageing, of the Universe. If this hypothesis were correct, then it should be possible to evolve one particular theory on the basis of which definite predictions could be made of the ratios of different forces at different times.

Despite the arguments presented above, it must be emphasized that for the present the idea of variable gravity, of G declining in proportion to the expansion of the Universe, is not generally accepted.

Unified forces?

We recognize today the existence of four forces of nature, these being the strong and weak nuclear interactions, discovered in the twentieth century, and the gravitational and electromagnetic forces, which, in some shape or form, have been known for a long time. The relative strengths of these forces are given in Table 3.

The unification of forces has been a continuing theme in physics during the past century. As we saw in Chapter 4, during the nineteenth century Maxwell showed that the electric and magnetic forces were manifestations of one common force, the electromagnetic force. Einstein tried unsuccessfully to develop a "unified field theory" which would bring together the forces of gravity and electromagnetism in one all-embracing theory, but electromagnetic theory went on to achieve great successes when it was adapted to conform with quantum mechanical principles in the theory of quantum electrodynamics.

In 1967 S. Weinberg of Harvard and A. Salam of Imperial College, London, and the International Centre for Theoretical Physics in

Table 3 The Forces of Nature

Force	Strength of force (between two protons)	Range	Force carrier
Strong nuclear interaction	1	10^{-15} metres	gluon* meson†
Electromagnetic interaction	10^{-2}	long	photon
Weak nuclear interaction	10^{-5}	10^{-17} metres	intermediate vector boson
Gravitation‡	10^{-39}	long	graviton

* The gluon is the force-carrying particle exchanged between quarks which binds together bunches of quarks making up baryons (e.g., protons and neutrons) and mesons.

† The strong nuclear interaction *between* hadrons (baryons and mesons) is communicated by the exchange of mesons, but mesons themselves are composed of quarks bound together by gluons.

‡ Gravitation is the only force which is always attractive.

Trieste, Italy, independently developed a theory which unified the electromagnetic force and the weak nuclear interaction, proposing that they are manifestations of the same force. This assertion has been confirmed by a number of experimental results. For example, among the predictions of the theory is that electrons interacting with protons will show a slight preference for "right-handed" interactions over "left-handed" ones, so that the rate of interactions involving right-hand spinning electrons should differ from the rate involving left-hand spinning electrons. Such effects have been observed, and they demonstrate that in this situation nature violates parity (the "law" that there is no distinction between right-handed and left-handed properties of quantities and interactions in nature).

Since the strong nuclear interaction is described in terms of a similar type of theory, there is optimism that it, too, will soon be drawn into the fold, leaving only gravitation out on a limb.

GUT theories

Theories which attempt to link together the strong, weak and electromagnetic interactions are known as grand unified theories, a term which is abbreviated, somewhat irreverently, to GUTs. Several such theories are under investigation, the basic premise being that one original force has evolved into separate manifestations. The analysis of GUTs developed by M. Yoshimura of Tohoku University, Japan, suggests a natural process by which the dominance of matter over antimatter may have emerged in the Big Bang, and leads to the prediction that the ratio of the number of photons to the number of protons in the Universe should be about 10^9:1, a figure which lies within the range of current estimates.

As we noted earlier, "heavy" particles such as protons, antiprotons, neutrons and antineutrons are known as *baryons*, and associated with them is a quantity called *baryon number* which can take the values +1 (e.g., for protons) or −1 (e.g., for antiprotons). Lighter particles, such as electrons, carry zero baryon number. One of the best attested laws of today's physics is the conservation of baryon number: in any interaction involving baryons, the net baryon number should be preserved; for example, if three particles of +1, +1 and −1 baryon number were involved in a mutual interaction, whatever emerged from that process would have a net baryon number of +1. A consequence of this law is that protons cannot decay into lighter particles and radiation, because in so doing they would destroy baryon number. It has therefore been assumed that protons are stable for ever.

But black holes would appear to violate this rule, particularly if they do indeed radiate equal numbers of particles and antiparticles as the Hawking process indicates. For a black hole formed entirely from particles (no antiparticles present in the material from which it formed) would be "processed" into equal numbers of particles and antiparticles. Thus a mass of material initially with a very large positive baryon number would end with a net baryon number of zero.

The half-life of a proton—i.e., the time required for 50% of a sample of protons to decay—must be greater than 10^{30} years for the occasional break-up of one not to have been noticed by now. Current GUTs suggest that the half-life may be in the neighbourhood of 10^{34} years, and an experiment is under way to try to detect the predicted decay. A joint group of researchers from the Universities of California and Michigan and from the Bookhaven National Laboratory have installed a huge plastic bag containing some 10,000 tonnes of ultra-pure water at a depth of over 600m in a salt mine near Cleveland. Surrounded by detectors to measure any radiation emitted by proton decay, the bag is located so far underground to minimize the number of spurious events arising from, for example, cosmic rays. Even in so great a mass of water, the most the experimenters can hope to detect is a few hundred proton disintegrations per year if current GUTs are correct. If the experiment produces positive results, it will provide powerful confirmation that physicists are on the right lines in their attempts to unify three of the four forces—and it will enhance the conviction that gravity, too, must eventually be made to fit into a single unified framework.

Quantum gravity

The weak and strong nuclear interactions and the electromagnetic force can, separately or in unified form, be described in terms of quantum field theories; i.e., theories which incorporate the principles of Special Relativity and quantum mechanics: Special Relativity contributes the equivalence of mass and energy; quantum mechanics injects the ideas of discrete units of energy and the uncertainty principle. Modern quantum field theories of these three forces are of a type known as "gauge" theories, the forces themselves being described in terms of the exchange of particles between interacting particles. The particles exchanged are known as "virtual" particles because they cannot be seen directly, and exist only fleetingly.

As an analogy (which should not be taken too far) we can visualize

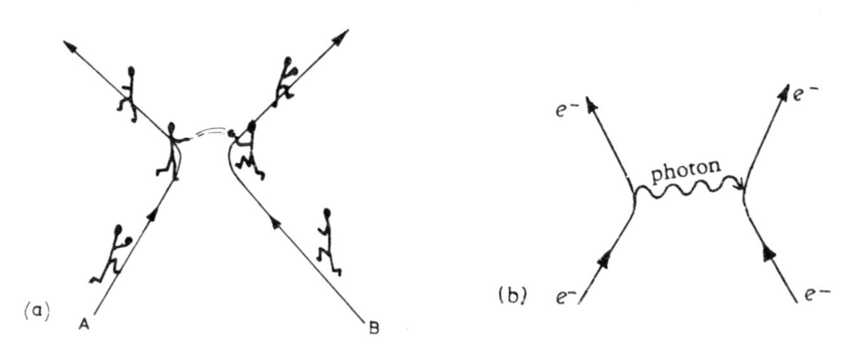

Fig. 51 *Feynman diagrams.* (*a*) The modern quantum-field theories of the weak and strong nuclear interactions and the electromagnetic interaction envisage forces being communicated by the exchange of "virtual" particles. The analogy here is of a pair of athletes whose converging courses are made to diverge by the action of one throwing a heavy medicine ball to the other. Similarly (*b*), in a Feynman diagram of the electromagnetic interaction, two electrons of like charge (e^-) approach and are repelled by the exchange of a photon, which materializes for an infinitesimal period to communicate the interaction.

how the process operates by thinking of two athletes running on converging paths, one carrying a heavy medicine ball (fig. 51). When they get close enough he throws his ball to the other, and as a result the athletes are pushed away from each other. For example, the electromagnetic interaction between two particles of like charge may be visualized in this kind of way: as the electrons, say, approach each other they exchange a particle—in this case, a massless photon—and the result of the interaction is mutual repulsion. A process of this kind can be described in terms of a Feynman diagram. The virtual photon exchanged exists for only a microscopic interval of time; that it exists at all follows from the uncertainty principle, whereby it is impossible to know the precise amount of energy contained in a microscopic system for an indefinitely short period of time. Consequently, quantum mechanics allows a particle of given energy to materialize for a brief instant; the higher the energy of the particle, the shorter the interval of time for which it can exist. We saw this kind of process in action when we considered the way in which particles might be created in the vicinity of a black hole (Chapter 9).

Looked at from this point of view, long-range forces such as the electromagnetic force can be communicated only by massless particles, because these are the only ones which can exist for long enough to act over large distances. Short-range forces (the weak and strong nuclear interactions) are communicated by the exchange of mass-possessing virtual particles which, because of their high energies, can exist for only extremely short times and so cannot be involved in interactions over large distances. The force-carrying particle of the electromagnetic interaction is the photon, while that of the weak interaction is known as the *intermediate vector boson* (of which there are thought to be three types, positive, negative and neutral) and is rather like the photon but with a finite mass. The strong interaction between nuclear particles is communicated by *mesons*, this interaction being a manifestation of the more fundamental force which binds together bunches of quarks to make up particles such as protons, neutrons and mesons; the force-carrier in this latter case is the *gluon*. Presumably, a quantum theory of gravitation can be expected to operate in the same kind of way, the interaction being communicated by massless particles called *gravitons*.

A distinguishing feature of different families of particles is the property of *spin*—we can visualize the particles as spinning on an axis, although we must bear in mind that subatomic and nuclear particles are not really like little billiard balls. Units of spin are such that familiar particles like electrons, protons and neutrons have spin values of $\frac{1}{2}$, and massless particles such as photons have spin $= 1$. The particles exchanged in the weak, strong and electromagnetic interactions each carry spin $= 1$, which results in repulsion between like particles (e.g., two electrons) and attraction between unlike particles (e.g., a proton and an electron). Gravitons are considered to be of spin $= 2$, since *all* interactions involving the exchange of spin-2 particles would result in attraction.

Although it is assumed that a quantum theory of gravity will involve the exchange of gravitons, there is as yet no theory which works. However, one promising development is *supergravity*, a theory devised in its first formulation in 1976 by D. A. Freedman, P. van Niewenhuizen and S. Ferrara and, independently, by S. Deser and B. Zumino. In supergravity, there is a single kind of particle, a *superparticle*, which can appear in various guises as any kind of force-carrying particle, or as a quark or lepton (a "light" particle, such as an electron), so linking the graviton to all other fields and particles. In this approach it is possible to start from the concept of a spin-2 graviton and build up a theory of gravitation whereby material particles are influenced, by the exchange of virtual gravitons, in accordance with Einstein's equations of General Relativity.

The long-range force of gravity is communicated by the exchange of massless gravitons of spin-2. Supergravity requires also that there exist massive spin $\frac{3}{2}$ particles, or *gravitinos*. The effects of the exchange of gravitinos will become apparent only at very short ranges where they should produce modifications to General Relativity.

A rough analogy for the behaviour of a superparticle can be obtained by thinking of it in terms of dice. Depending on how it is "rotated" a die assumes different values from 1 to 6. Likewise, by "rotating" a superparticle we can end up with all the particle types in the Universe. By progressive transformations one can go from a spin-2 graviton to a spin-$\frac{3}{2}$ gravitino, to spin-1 particles like photons, to spin-$\frac{1}{2}$ particles like electrons and protons, and finally to the spinless particles. The theory does not yet account for all the particles in the real world, and for their differing masses, but it does show considerable promise.

Further tests of General Relativity

While these various theoretical developments continue, the fact remains that General Relativity—a classical field theory, treating gravitation in terms of the curvature of space-time—is highly successful. In recent years experimentalists have been devising new and more stringent tests of its predictions compared to those of its rivals.

In order to be viable, any alternative theory of gravitation must satisfy the weak equivalence principle—that all bodies, whatever their mass or composition, experience the same acceleration in a given gravitational field. This has been confirmed to an accuracy of 1 part in 10^{12} by experiments such as those carried out by the Soviet physicists V. B. Braginsky and V. Panov. General Relativity satisfies the strong equivalence principle that all freely falling nonrotating laboratories are fully equivalent for the performance of experiments, wherever that laboratory may be located in the Universe, and at whatever time. Not all of the competing theories incorporate this requirement: according to some there may be "preferred frame" effects and/or "preferred location" effects. The former would imply that the motion of a laboratory compared to the frame of reference constituted by the galaxies would affect the results of gravitational experiments carried out inside that laboratory, while the latter would imply that the close proximity of a massive body could produce discernible effects inside a freely falling laboratory, contrary to the strong equivalence principle. Tests have already been able to set limits on such effects.

Another aspect in which competing theories differ is in the amount of space-time curvature which they predict will be produced by a given quantity of matter (we have already seen how the Brans–Dicke theory predicted a different value for the advance of the perihelion of Mercury, this effect being due to the degree of space-time curvature present in the vicinity of the Sun). The degree of bending of light by the Sun is one test of the amount of space-time curvature present. Another effect, not predicted by Einstein but discovered to be a consequence of General Relativity by I. I. Shapiro in 1964, is the time delay which occurs in radar echoes from a target when the beam passes close to the edge of the Sun; this, too, depends upon the degree of space-time curvature. Experiments carried out with increasing precision since 1968 by Shapiro and others, using radar reflections from planets and signals relayed by the Viking orbiters circling Mars, agree very closely with the predictions of General Relativity.

Light bending and radar time-delay measurements to date confirm that the amount of space-time curvature in the vicinity of the Sun agrees with the predictions of Einstein's theory to well within 1%. This means that any viable alternative theory has to conform in this respect with General Relativity to within that degree of precision.

General Relativity has survived all tests to which it has been subjected so far but, although the tests place limitations on alternative theories, and have already eliminated some, the fact remains that many such theories are viable, and more searching tests are required to seek out the weaknesses of all theories of gravitation.

Gravitational waves

The most promising direction of investigation at the moment lies in the search for gravitational waves. If an electrical charge is oscillated to and fro it will emit electromagnetic waves, as Maxwell predicted and Hertz confirmed. As General Relativity is a field theory similar to electromagnetic theory (in its original guise at least), it seems reasonable to suppose that oscillating masses will produce wave disturbances in the gravitational field—i.e., gravitational waves—which should travel through space at the speed of light: if the Sun were to be annihilated it would be 8.3 minutes before the Earth ceased to "feel" the gravitational influence of the Sun.

Although all viable theories of gravitation predict the existence of gravitational waves, they ascribe different properties to them; in some theories the speed of propagation could differ from that of light. Detection of gravitational waves and the analysis of their properties would provide a crucial test of gravitational theories.

The expected properties of gravitational waves, according to General Relativity, differ in a number of respects from those of electromagnetic waves; for example, because gravitation is so much weaker than the electromagnetic force, they must be very much weaker than electromagnetic waves. They are expected to affect matter differently, too. If an electromagnetic wave were to pass through a cloud of electrons, say, it would cause them all to move up and down in unison, just as water particles move when a water wave passes by, but a gravitational wave passing through would distort the cloud in the way shown in fig. 52. For example, if a gravitational wave were

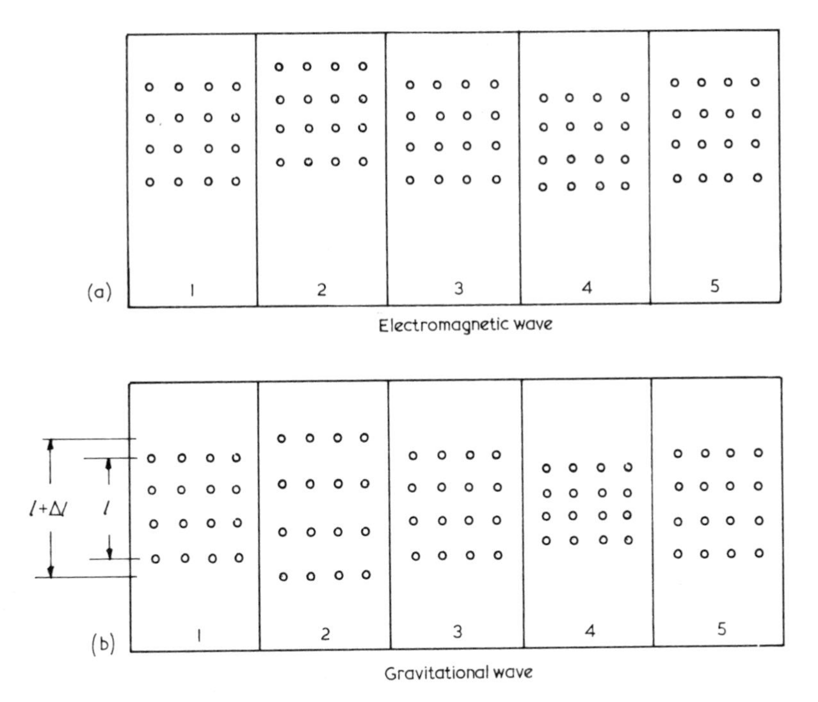

Fig. 52 *The effects of electromagnetic and gravitational waves passing through a cloud of electrons.* (*a*) As an electromagnetic wave passes through, the electrons move up and down in step with each other. Frame 1 depicts the situation at the beginning of one cycle of the wave; frames 2, 3, and 4 show the situation after $\frac{1}{4}$, $\frac{1}{2}$ and $\frac{3}{4}$ of a cycle. (*b*) A gravitational wave causes the electrons to move with respect to each other. After $\frac{1}{4}$ cycle (frame 2) they are stretched apart in the vertical plane and squeezed together in the horizontal plane; after $\frac{1}{2}$ cycle they return to their original disposition prior to being stretched apart in the horizontal direction and squeezed in the vertical direction at $\frac{3}{4}$ cycle (frame 4).

to pass perpendicularly through this page, the page would become longer and narrower, and then shorter and fatter.

All accelerating masses are expected to emit gravitational waves: if you jumped up and down you would emit gravity waves, although far too weakly for there to be any prospect of measuring them! Bodies in orbit around each other are continually accelerating; but, if we take the case of a typical binary consisting of two stars of solar mass separated by 1 astronomical unit, the power output in gravitational waves would be less than one hundred-millionth-millionth of the output in electromagnetic radiation from the stars concerned. Clearly we have to look at more exotic sources to have hopes of detecting substantial amounts of gravitational radiation. Close binaries involving white dwarfs, neutron stars or black holes offer progressively better potential sources, and other good prospects include stellar collapse to form black holes, supernovae—particularly where the explosion is asymmetric—and black-hole events in globular clusters, galactic nuclei and quasars.

Table 4 Estimates of gravitational radiation from plausible sources

Source	Typical frequency (Hertz)	Typical wavelength (metres)	Typical $\Delta l/l$*	Possible event rate
Stellar collapse: supernovae, black hole formation, in our Galaxy	10^2–10^5	3×10^6– 3×10^3	10^{-18}–10^{-17}	1 in 30 years
in Virgo cluster	10^2–10^5	3×10^6– 3×10^3	10^{-20}–10^{-21}	about 10/year
Black hole events† globular clusters: black holes of 10^2–$10^4 M_\odot$	1–10^2	3×10^8– 3×10^6	10^{-20}	up to 1/month
Galactic nuclei: black holes of 10^6–$10^{10} M_\odot$	10^{-6}–10^{-2}	3×10^{14}– 3×10^{10}	10^{-16}	up to 50/year
Binaries in our Galaxy	10^{-5}–10^{-2}	3×10^{13}– 3×10^{10}	10^{-19}–10^{-22}	continuous
specific example, Iota Bootis	8.6×10^{-5}	3.5×10^{12}	5×10^{-21}	continuous

* The present detectivity of gravity wave detectors is about $\Delta l/l = 10^{-16}$.
† Figures relate to events out to several billion light years' range.

Data based primarily on figures quoted by D. H. Douglass and V. B. Braginsky in *General Relativity—An Einstein Centenary Survey*, edited by S. W. Hawking and W. Israel, Cambridge University Press, 1979.

A useful way of characterizing the strength of an incoming gravitational wave is to examine the amount (Δl) by which the separation (l) of a pair of particles changes when a wave impinges at right angles to the line joining them, since over moderate distances the quantity $\Delta l/l$ should remain the same whatever the separation of the particles.

For realistic events, these changes are minute: the collapse of a star in our Galaxy should render a value for $\Delta l/l$ of 1 part in 10^{17} to 1 part in 10^{19}—and such events probably occur on average only once every 30 years or so at most. A pair of particles one metre apart would be shifted by such an event through a distance smaller than the size of a nuclear particle; even a pair of artificial satellites a million kilometres apart would be displaced by only about 100 times the diameter of a hydrogen atom. The displacements expected to be produced by various possible sources of gravitational waves are indicated in Table 4.

Possibly the most hopeful area of investigation would be events associated with supermassive black holes in galactic nuclei and quasars; estimates by K. Thorne and Braginsky suggest there may be as many as 50 such events per year, producing a $\Delta l/l$ of about 10^{-16}.

Various kinds of gravity-wave detectors are possible. Possibly the simplest would be a metal bar whose ends would endeavour to oscillate to and fro as a gravitational wave passed through: by measuring the strains set up in the bar the gravitational wave could in principle be detected. A given size and mass of bar will respond most favourably to waves of a particular frequency (i.e., it will resonate)—generally speaking the frequencies of gravitational waves are expected to be very low (from 10^{-3}Hz to about 10^{5}Hz) with correspondingly long wavelengths (3×10^{11}m to 3000m, respectively).

The first full-scale programme to try to detect gravitational waves was undertaken by Professor J. Weber of the University of Maryland who, together with his students, built a detector from a solid cylinder of aluminium about 1.5m long and weighing several tonnes. To it were attached extremely sensitive strain gauges capable of detecting minute distortions in the cylinder. Unfortunately, there are many local events—even someone walking past—which could cause the sensors to respond. Accordingly, Weber built a second detector at the Argonne National Laboratory near Chicago, some 800 kilometres distant, and argued that the same spurious events could not be recorded simultaneously on both detectors; if both detectors recorded an event at the same instant, he argued, they must be responding to genuine gravitational waves.

In 1970 he announced the detection of gravity waves, and located the source of these as lying somewhere in the direction of the galactic centre. A problem arose: for his primitive detector to be able to pick up events at that range, the energy being released from the galactic centre in the form of gravitational waves would have to be stupendous—equivalent to several thousand supernovae, stellar collapses, or similar events such as the swallowing by a central black hole of several thousand solar masses, *per annum*. The results seemed highly implausible, although there was the possibility that some kind of gravitational lensing effect might concentrate the radiation into the plane of the Galaxy, rather than permitting it to move outwards uniformly in every direction, so that the power output necessary to produce his results would not need to be so great.

Other groups in various centres throughout the world have built Weber-type detectors, some more sensitive than his original device, but all have failed conspicuously to detect any gravity waves. The conclusion now seems to be that Weber did not, after all, detect genuine gravitational waves, although a satisfactory alternative explanation of his results is not readily apparent.

If the figures in Table 4 are at all accurate, then plausible levels of gravitational radiation lie below the capabilities of existing detectors. Nevertheless, work is proceeding with enthusiasm at a number of centres throughout the world, and there is real optimism that within the next ten years or so devices will achieve the sensitivity required for the clear-cut detection of gravitational waves. It would be singularly appropriate if this discovery were to come in the decade which contains the 100th anniversary of the Michelson-Morley experiment and the 300th anniversary of the publication of Newton's *Principia.*

Other types of detectors currently under investigation include free-mass detectors, where two or more masses are free to move in space relative to each other (clearly experiments with *completely* free masses could take place only in space; "nearly free" masses can be achieved in the laboratory with pendulums), and resonant detectors, the simplest form of which would be two masses linked by a spring; like any spring, such a detector would have a particular resonant frequency.

Background noise is a problem, and this can best be reduced by increasing the masses of the detectors and reducing the temperature of the experimental apparatus to as near to Absolute Zero as possible. Giant single crystals of sapphire or silicon show considerable promise as detectors.

Laser-ranging detectors also are being regarded with optimism. In essence these rely upon measuring the displacement of two mirrors with respect to some reference mass. Folding the light-path by means of multiple reflections between the mirrors allows their effective separation, and therefore the effective length of the detector, to be greatly increased. With suitably coated mirrors up to 300 consecutive reflections can be achieved. By this means, the change in path length caused by the gravitational waves could be magnified several hundred times, and with this technique it may be possible to simulate a gravitational-wave antenna as much as a few hundred kilometres long. Systems of this type are being developed by, among others, Professor R. Drever's group at Glasgow University.

Other possibilities include the very careful monitoring of the motion of interplanetary spaceprobes by studying the Doppler effect on their transmitted signals, and—looking further ahead—laser tracking between a pair of widely separated spacecraft may give sufficient sensitivity to pick up even the weak gravitational waves from normal binary systems.

If gravitational waves can be detected directly, a whole new branch of astronomy, probably just as revolutionary as radioastronomy or X-ray astronomy have turned out to be, will be opened up. Gravitational waves can penetrate depths of matter opaque to other radiations, allowing the study of phenomena associated with the interiors of supernovae, with black holes and galactic nuclei and—if the required sensitivity could ever be achieved—of events associated with the era of galaxy formation; perhaps the early instants of the Big Bang itself would come into "view", allowing us to "feel" the rumble of the event and gain information relating to a time long before the release of the microwave background radiation. For the moment, we must await the development of the necessary instrumentation.

If we cannot yet detect gravitational waves directly, perhaps we can observe the side-effects of their emission. For example, in a close binary system the emission of energy in the form of gravitational waves should cause the stars slowly to spiral together. The pulsar PSR 1913+16, which was detected in 1974 by R. Hulse and J. Taylor using the giant Arecibo radio dish, turned out to be a member of a binary system with a period of 8 hours, the pulsar itself having a period of 0.059 seconds. The discovery aroused great interest. On the one hand, General Relativity could be used to assess the properties of the system—the combined masses of the bodies involved was estimated at 2.83 solar masses on the basis of the enormous (by comparison with Mercury) perihelion advance of 4.22 degrees per year. On the other hand, the binary system could be used as a laboratory in which to test General Relativity against its competitors.

The type of gravitational radiation predicted by General Relativity is known as quadrupole radiation, and the change in orbital period that should be brought about by its emission has been calculated to be about 2 parts in 10^9 per year. Another type of gravitational radiation—dipole radiation—is predicted by some competing theories, and should lead to a decrease in period of about 3 parts in 10^7. Results reported by Taylor in late 1978 indicated that the period is declining by about 3.6 parts in 10^9 per annum, a figure more in accord with General Relativity than with several of its competitors.

Do these observations represent the successful detection of the effects of the emission of gravitational radiation and yet another triumph for General Relativity? It is too soon to be sure, for there are other possible explanations of the binary pulsar's behaviour. Tides raised in the companion star would significantly affect the period and evolution of the orbit unless the second star, too, were a compact body like a neutron star or black hole. Optical observations made by P. Crane, J. E. Nelson and J. A. Tyson at Kitt Peak National Observatory indicate the presence of a very faint star close to the pulsar's position: if it is the pulsar's companion, then it is too bright to be a white dwarf (or neutron star or black hole). It may be a helium star—a star which has lost its outer envelope of hydrogen—on which significant tides would be raised, completely clouding the issue so far as relativistic effects are concerned.

The matter is unresolved at the time of writing, but the binary pulsar seems to be such a godsend to relativists that it would be a tragedy if it turned out not to be a suitable testing ground for General Relativity after all.

Gravity today... and tomorrow

General Relativity still reigns supreme. However, gravitational theory is moving rapidly out of the realms of pure mathematical physics into the hard arena of experiment. In the years ahead we may expect to see increased precision in the "classical" Solar-System tests—the gravitational red-shift/time dilation, perihelion advance, deflection of light and signal delay. With the binary pulsar, although there are doubts about its true nature, hopes are high that the testing ground of General Relativity has already moved out to stellar systems; and there is a strong expectation that gravitational waves will be detected before long, opening up new astronomical horizons and permitting the making of crucial tests of the relative merits of General Relativity and its competitors.

On the one hand, sophisticated laboratory and space-based experiments are under consideration; on the other, a discovery such as the unequivocal detection of black holes would do much to enhance our understanding of gravity, space, time and the Universe. Exploding primordial black holes are the joker in the pack: there is considerable debate as to whether or not they exist, but if they do, and if they can be observed in the act of exploding, then positive links between quantum mechanics and gravitation will have been established, and theories of particle physics will be subjected to tests unattainable in any terrestrial laboratory.

On the theoretical side, with the unification of the weak nuclear and electromagnetic interactions already apparently achieved, and with experiments under way to test the new generation of GUTs, which attempt to unify the weak, electromagnetic and strong interactions, there is an air of optimism that the day may not be too far distant when all four of the forces of nature are brought together under the same umbrella. Perhaps supergravity will prove to be the answer—we cannot say at present. Nor can we predict what effects a new quantum theory of gravitation may have on our view of the Universe, its origin and evolution, or on such vexed questions as the nature of singularities inside black holes.

The study of gravity, black holes and the Universe has become a central issue in the onward march of Man's attempts to comprehend the world around him and the forces which shape its destiny. We have come a long way from the ideas of Aristotle, Galileo, Newton and even Einstein himself. The coming decades may see new revolutions just as great as those which have already shaped our comprehension of the cosmos.

Bibliography

Although there is a wide variety of books written at a range of different levels on the various aspects of the subject matter of this one, it must be borne in mind that gravitation, black holes and cosmology are areas of intense research activity, and in order to keep up to date it is necessary to consult the current journals. Periodicals which contain up-to-date news and popular review articles include *New Scientist* (UK, weekly), *Scientific American* (USA, monthly), and *Sky and Telescope* (USA, monthly). Technical journals which also include news and review articles include: *Nature* (UK, weekly), and *Science* (USA, weekly). Specialist technical journals of particular relevance include *Astronomy and Astrophysics*, *Astrophysical Journal*, *Comments on Astrophysics* (very useful reviews), and *Journal of General Relativity and Gravitation.* A most useful annual publication is *Annual Review of Astronomy and Astrophysics*, edited by G. Burbidge, D. Layzer and J. G. Phillips, Annual Reviews Inc., California.

A personal selection of books is given below. Many of those listed are at a significantly higher technical level than the present book. With the exception of the general and historical works, those at a nonmathematical and more popular level are indicated with an asterisk.

General astronomy

Meadows, A. J.: *Stellar Evolution* (3rd edn.), Pergamon, Oxford, 1978.

Mitton, Simon (ed.): *The Cambridge Encyclopaedia of Astronomy*, Cape, London, 1977.

Nicolson, Iain: *Astronomy—A Dictionary of Space and the Universe*, Arrow, London, 1977.

Pasachoff, J. M., & Kutner, M. L.: *University Astronomy*, W. B. Saunders, Philadelphia, London, Toronto, 1978. A wide-ranging modern text with very little mathematical content; a shorter, nonmathematical, version is entitled *Astronomy Now.*

Ronan, Colin (ed.): *Encyclopaedia of Astronomy*, Hamlyn, London, 1979.

History of force, gravitation and cosmology

Berry, Arthur: *A Short History of Astronomy*, Dover, New York, 1961. This is a reprint of a work originally published in 1898.

Cohen, I. Bernard: *The Birth of a New Physics*, Heinemann, London, 1961. A concise account of the emergence of Newtonian mechanics and gravitation.

Grant, E.: *Physical Science in the Middle Ages*, Wiley, New York, Chichester, 1971.

Koyré, A.: *From the Closed World to the Infinite Universe*, Johns Hopkins Press, Baltimore, 1957.

Ronan, Colin: *Galileo*, Weidenfeld and Nicolson, London, 1974. A very readable account of the life, work and times of Galileo.

Shea, W. R.: *Galileo's Intellectual Revolution*, Macmillan, London, 1972.

Westfall, R. S.: *Force in Newton's Physics*, Wiley, New York, Chichester, 1971.

Modern concepts of force and particles

*Calder, Nigel: *The Key to the Universe*, British Broadcasting Corporation, London, 1977. Excitingly written introduction to the modern physics of force.

*Davies, P. C. W.: *The Forces of Nature*, Cambridge University Press, Cambridge, 1979. Thorough, yet nonmathematical.

Relativity, gravitation, black holes and cosmology

*Bergmann, P. G.: *The Riddle of Gravitation*, Murray, London, 1969. Although dated in some aspects, this remains a clear and lucid discussion of the subject.

Berry, M.: *Principles of Cosmology and Gravitation*, Cambridge University Press, Cambridge, 1976. Clear, concise and of moderate technical difficulty.

*Einstein, Albert: *Relativity—The Special and the General Theory*, Methuen, London, 1920; paperback edition, 1960. A concise introduction "straight from the horse's mouth".

Hawking, S. W., & Israel, W. (eds.): *General Relativity – An Einstein Centenary Survey*, Cambridge University Press, Cambridge, 1979. An extensive collection of up-to-date review articles, many at a high level of difficulty, by leading researchers in the field. Nevertheless, several of the reviews are accessible to nonspecialist readers. An outstanding volume.

Hazard, C., and Mitton, Simon: *Active Galactic Nuclei*, Cambridge University Press, Cambridge, 1979. Although highly technical in parts, this book includes material accessible to the nonspecialist and provides a useful reference source in a rapidly developing area of research.

*Hoyle, Sir Fred: *Ten Faces of the Universe*, Freeman, San Francisco, 1977. A collection of fascinating and provocative essays.

*John, L. (ed.): *Cosmology Now*, British Broadcasting Corporation, London, 1973. A collection of highly readable articles based on a series of broadcasts.

⋆Kaufmann, W. J.: *The Cosmic Frontiers of General Relativity*, Little, Brown, Boston, 1977; Penguin, London, 1979. A clear, readable and copiously illustrated account of Relativity and of black holes. Pays particular attention to space-time diagrams and to depicting what would be seen by observers entering black holes.

Misner, C. W., Thorne, K. S., & Wheeler, J. A.: *Gravitation*, Freeman, San Francisco, 1974. An outstanding graduate-level text.

⋆Narlikar, J. V.: *The Structure of the Universe*, Oxford University Press, Oxford, London, 1977. Interesting discussion of aspects of cosmology, gravitation, time and astrophysics by one of the originators of the Hoyle–Narlikar theory.

Rindler, W.: *Essential Relativity* (revised edn.), Van Nostrand Reinhold, New York, Wokingham, 1977. One of the clearest semitechnical expositions of the Special and General theories.

Sciama, D. W.: *Modern Cosmology*, Cambridge University Press, Cambridge, 1971. Readable and clear introduction to cosmology and relevant aspects of astrophysics, with only a little mathematics.

Wald, R. M.: *Space, Time and Gravity*, University of Chicago Press, Chicago, 1977. Concise account, based on ten public lectures.

Weinberg, Steven: *Gravitation and Cosmology: Principles and Applications of the General Theory of Relativity*, Wiley, New York, Chichester, 1972. Advanced general treatise.

⋆Weinberg, Steven: *The First Three Minutes*, Basic Books, New York, 1977; Deutsch, London, 1977. *The* outstandingly lucid and readable discussion of Big-Bang cosmology.

Index

INDEX